THE
VISITOR'S
REPORT

The End ... And New Beginning
of the Human Race

By
Robert Vanderzee

I loved the premise of this book! "The Visitor's Report" by Robert Vanderzee takes all of the current and new theories about the origins and make of the Universe, Earth, and even our own origin as a species, and blends them in with UFO sightings, alien theories, religion, history and science, to create a thought-provoking thesis on how things really went down.
I was hooked from the beginning. . .

The persons portrayed in this book are fictitious. No similarity to actual persons, living or dead, is intended or should be inferred.

Copyright © 2013 Robert Vanderzee
All rights reserved.
ISBN: 148496389X
ISBN 13: 9781484963890
Library of Congress Control Number: 2013909615
CreateSpace Independent Publishing Platform
North Charleston, South Carolina

Acknowledgment

I want to thank the members of the Palm City Word Weavers authors group for their contributions to this work. Their thorough, concise, and professional comments contributed hugely to whatever success this work may have. My sincere thanks to all, and particularly to Leona DeRosa Bodie, who organized the PCWW and continues to bring the best out of each of us who attend her regular meetings.

DEDICATION

This book is dedicated to the late Robert Moeller, suspected by his family and a very select few of his earthly friends to be none other than VAATU-9 himself.

Short Biography of Author

Robert Vanderzee is a graduate of the University of Michigan and a commercial pilot. He worked as a mechanical engineer, living at various times in Akron, Ohio; London, England; Bethune, France; and Carlisle, Pennsylvania. He loves boating, flying airplanes, skiing, photography, and writing books. He now lives in Port St. Lucie, Florida.

PREFACE

I have approached the subject of this work as a mechanical project engineer would an assignment to produce a big new machine. After all, I was such an engineer for over forty years, working for various engineering firms. This is how I think about great works of engineering.

And so, like a hammer believing every problem is a nail, I began my search for an answer to life's many unanswered questions—with the assumption that this magnificent edifice we call our universe is a project.

All of mankind's grand edifices—cathedrals, skyscrapers, bridges—are the result of engineering projects. No structure of any significance ever arose as a lone individual's achievement. So I see the universe as having been conceived and produced by a group, a vast engineering committee of designers, project engineers, managers, technicians, secretaries, and financial analysts, who toil in their laboratories and design facilities to achieve their goal or master plan.

Projects all begin with an idea. Then they are defined on paper so that others can be convinced the idea is a good one and should be financed. Once funds are available, committees are formed, tests are made, drawings drawn, and mechanical and electrical parts purchased and manufactured. Finally the object of the project is completed. But before it can be released for public use, it must be tested, and tested, and tested again. Every flaw in the product must be detected and eliminated.

All of this takes time, mental stress, anguish, personnel selections and dismissals, interminable progress meetings, and informed and uninformed questions from superiors, the answers to which must be researched and crafted so as to be sufficiently plausible to gain approval. Project engineering is a long, arduous trek, but by the end it is worth all the effort. Looking back at successful projects, all difficulties are forgotten. Successful project completion is a work of joy.

Individuals producing anything approaching the perfections, tolerances, and beauty of our elegant universe would have to be pretty damn smart—a lot smarter than any of us humans. Their product, of course, has flaws—in a way that makes them seem human. But like all project engineers, when they find flaws they search for ways to correct them.

In the pages to follow, I predict discoveries that will occur in "the near term." As I put these words to paper, I am over eighty years old and vividly recall many stories my seventy-year-old parents told me of their childhood. Thus for me, and anyone my age, "the near term" can extend for a hundred and fifty years. So, for the purposes of the story to follow, I feel I am safe in calling the year 2030, and later in the story, 2090, the near term. But please do not hold me to these dates. The predictions I make could easily be realized a century or more from now. But they will happen.

One last thought before I release you, dear reader, to rummage among my thoughts. Most readers think of our maker as "God," with a capital "G." But no one individual could have produced what we see around us. It must have been a group effort, one I think of as having been accomplished by a "committee." So if our committee is to replace God in the pages to follow, it is appropriate to capitalize the word. And like us humans, my Committee is flawed, capable of producing grand triumphs—and awful mistakes.

—*Robert Vanderzee*

INTRODUCTION

It is not unusual for revolutionary change in the thinking
of man to occur in the first thirty years of a century. Take
for example the nineteenth century. On 21 April, 1820,
Hans Christian Orsted, a Danish physicist, noticed during
a lecture he was giving to students that a copper wire, when
disconnected from a battery, deflected a nearby compass
needle. His intense study of the phenomenon after the lecture
provided the clues that led James Clerk Maxwell to his unified
model of electromagnetism. Thus began the exploitation
of electromagnetism otherwise known as electricity. The
dramatically new world you see around you today, with its
dynamos, electric motors, lights, and television, is the result.

In the first thirty years of the twentieth century, Einstein
produced his general theory of relativity, and eminent
physicists Niels Bohr, John von Newmann, Paul Dirac,
Werner Heisenberg, et al., arrived at their theories of quantum
mechanics. These ideas fundamentally redefined reality, and our
world was never the same. X-rays, radar, flights to the moon,

the global positioning system (GPS), cell phones, you name it—all stem from those early advances.

Now we are well into the first thirty years of our new twenty-first century. New ideas for fundamental change of our world are overdue. Where are these ideas? Could it be they reside on the pages to follow?

When faced with competing hypotheses, select the one that makes the fewest assumptions.
—Occam's Razor, est. 1852; named after
William of Ockham (c.1285-1349)

PART I

CHAPTER 1

QUESTIONS

There are three scientific puzzlements that, when solved, will allow mankind to discover its dreadful fate. Physicists and mathematicians vigorously pursue two of these perplexities. Scientists refuse even to discuss the third. All three will eventually find universal acceptance and convulse the world's population in a way not seen since the arrival of modern man here on Earth as the last ice age was ending, 11,500 years ago. They will define reality and the future in store for the human race.

There are many questions not adequately answered by science or by religion: Where among the some three billion DNA base-pairs of the human genome are the defects—the deficiencies—that allow an otherwise normal human being to strap explosives to his or her chest and seek out the largest crowd possible in which to detonate it? Why would God allow such awful things as the Thirty Years' War, the hideous extremes of the Spanish Inquisition, World Wars I and II, and the Holocaust to occur?

On a less horrendous level, why does man come in variants, i.e., Caucasian, Negroid, Oriental, Indian, Mayan, and so on, and how it is that these variants appeared on Earth at, cosmically speaking, roughly the same time? Other questions linger. *Homo erectus* appeared on Earth toward the end of the Pliocene epoch two million years ago and thrived for over a million years, only to die out when a superior species, *Homo sapiens*, appeared on the scene when the Epipaleolithic era dawned seventy thousand years ago. Why didn't these species coexist?

Nearly all accredited scientists believe in Darwin's theory of evolution, which says that, over immense time, brand new, more complex species suddenly appear and replicate. But the theory has huge holes in it. Holes you can drive a truck through. No one has explained, for example, in the absence of Darwin's proverbial "missing link," how that very, very first new-species infant, far more complex than its mother, arriving in its new world after billions of trials and errors, found the skill and luck to grow to adulthood, discover an identical member of the opposite sex, also having mutated after billions of trials and errors, to produce offspring. Evolutionists cannot explain this, yet they vigorously object to an alternative explanation: intelligent design.

Paleoanthropologists studying ancient Egyptian history can trace Egyptian civilization back only five thousand years, to small fifty- to two-hundred-member colonies and villages along the Nile. None can explain how these people could have survived in groups any smaller than fifty individuals. Did fifty evolve all at once? Nor can scientists find evidence of how they arrived there in the first place. One has only to see the huge trackless desert surrounding the Nile valley to find it inconceivable these people wandered there over this huge empty void, which had formed over 140,000 years before any sign of these people is found. These isolated communities of modern humans survived as hunter-gatherers for a thousand years before uniting and maturing into the Egyptian empire that stood for another three thousand. Where did they come from? Paleoanthropologists only guess.

Turning to other riddles, thirty to fifty sightings of UFOs are reported every month. Most are explained away, but about 5 percent have no conventional explanation. Recognized scientists will not touch this subject.

Here's another bafflement. Scientists have known for years that John von Neumann's quantum mechanics and Albert Einstein's theories are incompatible, yet they continue to use them to describe happenings in their separate spheres of interest, the world of the infinitesimal and the world of outer space. Physicists admit their formulae do not describe the entire real world, only a reality existing within narrow limits. They know a unifying theory must be found.

As you will soon see, I present answers to these riddles, comfortable that incontrovertible evidence will sooner or later confirm what I say. Some of you will think this is science fiction. But I assure you it is no more fiction than Isaac Newton's *Philosophiae Naturalis Principia Mathematica* (*The Mathematical Principles of Natural Philosophy*), published in 1687, some twenty-one years after Newton conceived the scientific concepts contained in it. Many philosophers of his day (now we call them scientists) thought it fiction when it was published, because his postulates—the very fundamentals on which his hypothesis was based—seemed fanciful to them and could not be proved. But within a generation it was recognized as a work of supreme genius.

During the last fifty years of the twentieth century, I collected, out of pure curiosity, a veritable library of diffuse facts, rumors, claims, faiths, sightings, and stories that might have a connection, no matter how tenuous, to the profound, unanswerable questions humans ask themselves on occasion. These bits and pieces of information pertain to such diverse subjects as: theology, stars in the sky, wars, gravity, UFOs, tiny building blocks of the universe called "strings," the speed of light, Einstein's theories, quantum mechanics, intelligent design, chaos theory, Darwin's theory of evolution, and many other curiosities. I filed this information away for a day when I would find time to arrange it into a coherent explanation of reality. Now that day has arrived.

In 1905 an obscure patent office employee in Berne, Switzerland, folded new facts, observations, and questions, into the scientific orthodoxy of his day to form an entirely new physics. I arrived at my hypothesis in much the same way. Certain my brain and thought processes will never be compared to Albert Einstein's, I nevertheless produced a story answering all the troubling questions of our day, and I bring it to you now.

To prepare my alternative to conventional thinking, I looked at all of Earth's most significant happenings over the centuries and rearranged them to exclude any reliance on, reference to, or suggestion of religion or Darwinian evolution. When theology and evolution are erased from one's thinking and are replaced with the implications stemming from string theory, UFOs, and intelligent design theory, the result is: THE VISITOR'S REPORT.

In the pages to follow, I will first prepare you for a voyage to our new reality and the future of mankind—and then I will take you there. For me this is easy, for I am what my superiors like to call an earth-monitor, someone, shall we say, who is looking out for your best interests. More about earth-monitors later.

Before I can go any further, however, you will need fully to understand just how fragile your own concept of *certitude* really is. Three other concepts, string theory, intelligent design, and unidentified flying objects, are explained in follow-on chapters. But you really must begin with the concept of *certainty*.

CHAPTER 2

CERTAINTY

You are absolutely certain of much that is around you. Doubtlessly certain. Anyone questioning your certainties causes your eyes to glaze over and look around for an excuse to avoid further discussion.

You live in a three-dimensional world: width, height, and depth. There are five, and only five, means of motorized transportation to choose from: boats, cars, trains, airplanes, or rockets. Your planet Earth revolves around its axis to give you day and night and ocean tides, and it orbits the sun to provide you four seasons.

Electricity lights your homes and offices, powers your tools, and operates your TV screens, cell phones, traffic lights, and the bar code scanners used at most every grocery store checkout counter. A jet takes you to Europe from New York in about six hours. Color television airs a multitude of movies, serials, news programs, and talking-head shows, enlightening you on every imaginable subject. Any of you can add hundreds of "certainties" to this list. These "certainties" form your world of the early twenty-first century—your reality, the world in which you live out your lives.

But a mere five hundred years ago—say, when Christopher Columbus visited America—none of these "certainties" existed. Anyone claiming the certainties of today might have been deemed a religious heretic and burned at the stake. In those days, transportation was limited to walking on foot, riding animals or animal-drawn vehicles, and those advanced engineering marvels of their day: sailboats. People then were utterly certain speed of travel was limited to whatever speed could be urged from animals or wind-driven boats.

The Greeks and the Church had taught for over twenty-five hundred years that the moon, sun, and stars revolved around Earth. This was accepted fact, and understandably so, because one had only to look skyward to see these heavenly bodies cross the sky as Earth apparently remained still.

So imagine how people reacted when first Johannes Kepler, and then Galileo, aided by his telescope, tried to explain that, sorry folks, the sun does not revolve around Earth, it only *appears* to. The Church had had it wrong for twenty-five centuries: the earth revolves on its own axis while orbiting the sun. All of this baffled the people of that day, because what Galileo was saying defied common sense, and, more importantly, it defied the teachings of the Catholic Church. The Church considered burning him at the stake, but it settled for keeping this strange man in house arrest—let's call it "protective custody"—for the remaining thirty years of his life. Galileo insisted to his dying day that the earth-centric teachings of the Catholic Church were wrong, and that the truth was just the opposite of what everyone apparently saw.

So how would Galileo have been greeted if, on top of all that he was teaching, he had promised the people of his day a time when vehicles would streak across the sky at enormous speeds, accomplishing in hours what it took Columbus months to do? What would they think if he had said that someday something called electricity would illuminate their homes, do their heavy labor, execute criminals, and enable television? How would those people have reacted if told their horses and wagons would be replaced by automobiles traveling smooth, paved roads at unheard-of speeds of sixty, eighty, and a hundred miles per hour? Witches were burned at the stake in those days for predicting far less. Would you really want to visit that world in your time machine and describe your modern world to them?

One needs only to go back to the year 1900 to find a world in which nearly every explanation of natural phenomena would be proved wrong. Physicists then were arguing man would never fly. Radio, television, computers, cell phones, iPods, a polio vaccine, antibiotics, and nuclear power had not been conceived. Flight to the moon was science fiction—utterly impossible. Scientists had no clues to plate tectonics, nanotechnology, the big bang theory, quantum mechanics, or the curved space of Einstein's general relativity. Within thirty years that old world had all changed.

If this old world certitude could vanish so quickly, how fragile are your certainties of today? Mankind's knowledge base is growing exponentially with the aid of the ubiquitous computer and Internet. As night follows day, I can assure you the certainties you are so comfortable with will vanish in the relative blink of an eye. They will be replaced by new beliefs, some that may be difficult and maybe even disturbing for you to imagine. Would it be out of line to predict that the long list of certainties you are so comfortable with will change as radically in the next thirty years as others had in the previous one hundred?

Centuries after Newton's day, Einstein recognized pesky flaws in Newtonian physics, and, based on a few fundamental assumptions of his own, went on to develop his own theories. Today, as in Einstein's day, and in Newton's day before him, physicists have accumulated more pesky flaws in and challenges to their favorite theories of how our world works. Examples are everywhere. Think of dark matter, dark energy, the "missing"

Higgs boson, UFOs, intelligent design, antimatter, black holes, string theory, and parallel universes, just to name a few.[1] Soon the Large Hadron Collider, a new high-energy proton collider that began operation near Geneva, Switzerland, in 2011, will reveal even more questions in the world of the small than it answers, just as the Hubble Space Telescope has done in the world of the large.

Today there is no end to the puzzles scientists may ponder, the answers to which will shatter the rock-solid certainties you have. But UFOs, string theory, and intelligent design are the three puzzles that should concern you most. Why? I will show you why. Let's examine them in detail.

---

1   Information about these and the many other challenges to modern-day particle physicists is easily found on the Internet.

CHAPTER 3

Unidentified Flying Objects

At 8:20 p.m. Mountain Standard Time, on March 13, 1997, witnesses in Paulden, Arizona, looked into the evening sky and saw a huge, V-shaped—some said "boomerang-shaped"—vehicle gliding silently toward the south. Witnesses in Chandler, Arizona, saw the vehicle as it approached, and soon thousands in the Phoenix area saw it as it appeared overhead, continuing its slow drift southward. By 9:00 p.m., it was over Tucson where witnesses there saw it hover, "fold its wings," then accelerate to enormous speed before disappearing from view. The incident was reported nationwide by the media.

Witnesses to this event, now considered by many as the largest mass sighting of a UFO in modern time, came from all walks of life and included physicists, airline pilots, police officers, engineers, and even the governor of Arizona, Fife Symington. Symington later called the event "otherworldly." One woman said the vehicle was so huge that if she had held an open newspaper up to the sky, it would have extended beyond the edges of the paper on all sides. All the witnesses agreed, and video footage confirmed, the craft was colored a dull gunmetal gray and was brightly lit at five, some said six, locations on its lower surface. Government spokesmen scoffed or else offered implausible explanations.

An hour after the craft disappeared over Tucson, Maryland Air National Guard, jets temporarily stationed at a nearby air base, scrambled into the air to drop flares a short distance from Phoenix[2] in what seemed an attempt to confuse the issue. Few were fooled.

Thirteen years earlier, on the evenings of December 26, 27, and 28, 1984, in what is known as the Rendlesham Forest Incident, US Air Force personnel spotted strange bright lights descending into a wooded area beyond the rear gate of the RAF Woodbridge Air Base in Suffolk, England, home to a nuclear stockpile and the largest collection of American aircraft outside the United States. Sergeant James Penniston and two others in the base security office were ordered to investigate the sighting.

They discovered a "craft of some sort" resting just inside the tree line beyond the base, emitting colored lights from inside its surface. They walked up to the vehicle, walked around it, looked under it, and found it to be triangular in shape, approximately nine feet on a side, and about six feet high, with a pyramidal dome accounting for perhaps three feet of its height. Its surface was seamless, very warm to the touch, and dull gray in color. It did not rest on landing gear but instead on a tripod of sorts extending from its underside. "The craft could not have been made by man," they reported. They made hurried notes in their notebook, including sketches of strange symbols etched into the surface skin. All the while they were in radio contact with the base and found that radio transmissions were hindered by static when they were near the vehicle.

---

2   This was a "diversionary tactical maneuver" later identified as Operation "Snowbird." Air traffic controllers confirmed the earlier sightings of UFOs occurred at a one-thousand-foot altitude in Class B restricted airspace.

Some time after the trio returned to the base, they watched the vehicle rise, make ninety-degree turns at high speed, and then disappear. Later over sixty base personnel saw, at one time or another, other such vehicles return to hover over the base ammunition dump, light it up with high-powered spotlights, and depart.

The next day, men returning to the craft's landing site discovered three round indentions about seven inches in diameter and one and one-half inches deep in the frozen ground where the vehicle had rested, each with radioactivity levels about eight times normal. James Penniston's official report, along with recordings of the radio transmissions, were sent up the military chain of command and never seen by them again. He kept his rough notes, which are revealed now.

In October 2009, the History Channel broadcast a two-hour, detailed review of unidentified flying object activity over the past fifty years. The broadcast included details of these and a number of other inexplicable UFO reports selected from the thousands that have been recorded since World War II. All of the witnesses interviewed for this History Channel presentation were highly intelligent and articulate.

As part of its presentation, the History Channel reported on a November 12, 2007, National Press Club meeting attended by scientists, airline pilots, air force generals, and government officials from around the world, including Arizona governor John Fife Symington. Speakers at this meeting described their experiences and many high-quality UFO sightings in 133 countries. They reported that the vehicles they saw were not aerodynamic, i.e., they did not fly in the sense that airplanes fly but instead must employ an unknown electromagnetic anti-gravity/inertia technology to control space-time in the vicinity of the craft, avoiding the need for massive supplies of fuel to move through our sky. (Already accredited scientists are probing ideas that dimensionless particles in the vacuum of outer space contribute to the inertial force that opposes acceleration of mass.[3])

No official comment from the US government has been made regarding this meeting or any other UFO sighting since "Project Blue Book" terminated in 1969. Most observers seem to agree that the reason official government sources either ridicule or do not comment on these sightings is 1) they would sound absurd to the public if they did comment, and 2) if UFOs do exist, they would represent a culture so advanced that human culture is powerless to explain or respond to it.

It was claimed in 2007 that all UFO sightings are anecdotal and can be explained as nothing more than prosaic natural phenomena. No official government investigation ever publicly concluded that UFOs are indisputably real physical objects, extraterrestrial in origin, or of concern to national defense. In none of the highly classified studies— such as Great Britain's Flying Saucer Working Party, Project Condign, the CIA-sponsored Robertson Panel, the US military investigation into the green fireballs from 1948 to 1951, and the Battelle Memorial Institute study for the USAF from 1952 to 1955 (Project Blue Book Special Report #14)—can one find flat-out statements of supernatural origins for these vehicles. Nevertheless, suggestions that the sightings could be extraterrestrial visits are embedded all through these and other reports.

In 1947 the US government established Project Sign to investigate reports of "flying saucers." Sign's personnel was left with the impression the subject would be taken seriously, on the grounds that UFOs may represent genuine aircraft whose origins are mysterious and possibly threatening to US security. Sign investigated earlier UFO reports, but it was the highly

---

3   Alfonso Rueda and Bernard Haisch, *Gravity and the Quantum Vacuum Inertia Hypothesis*, (Long Beach California State University: ManyOne Networks, 2005). Also J. Deardorff, B. Haisch, B. Maccabee, & H.E. Puthoff, *Inflation-Theory Implications for Extraterrestrial Visitation*, *JBIS* 58 (2005): 43-50.

publicized Chiles-Whitted UFO Encounter of July 24, 1948, that seemed to interest them most of all.

In that encounter, two experienced airline pilots claimed a torpedo-shaped object nearly collided with their commercial airplane. Sign personnel judged the report convincing and compelling, partly because the alleged object also closely matched the description of an independent sighting from The Hague a few days earlier. Despite the lack of physical evidence, some Sign personnel were swayed by this report and the reports of dozens of other aerial sightings they reviewed, many by scientists and military pilots.

Given that no evidence existed that either the US or the USSR had under development anything remotely like UFOs, Sign personnel gradually began considering an extraterrestrial origin for the objects. Pilots, engineers, and technical people have a "can-do" attitude and tend to regard unavailable technologies not as impossibilities but as challenges to be overcome. Rather than dismissing UFO reports out of hand, they began to consider how such objects might function. Their perspective is quite different from the attitude of many scientists, who characterize such concepts as impossible, unthinkable, or absurd. In 1948 Project Sign produced a highly classified opinion, an *Estimate of the Situation*, that the best UFO reports had an extraterrestrial explanation. Their opinion was submitted, traveled up the chain of command to then Chief of Staff, General Hoyt S. Vandenberg who rejected the Estimate and ordered all copies destroyed.

A top-secret Swedish military opinion given to the USAF in 1948 stated that some of their analysts believed that "ghost rockets" and later "flying saucers" reported there had extraterrestrial origins. In 1954 eminent German rocket scientist Hermann Oberth revealed that an internal West German government investigation he headed had arrived at an extraterrestrial conclusion, but this study was never made public. Classified internal reports by the Canadian Project Magnet in 1952 and 1953 also assigned high probability to extraterrestrial origins. However, neither Project Magnet nor later Canadian defense studies ever actually stated such a conclusion publicly.

The US National Security Council established another highly classified US study by the CIA's Office of Scientific Investigation in the latter half of 1952, which concluded UFOs were real physical objects of potential threat to national security. One memo stated, "The reports of incidents convince us that there is something going on that must have immediate attention. Sightings of unexplained objects at great altitudes and traveling at high speeds in the vicinity of major US defense installations are of such a nature that they are not attributable to natural phenomena or any known types of aerial vehicles." The matter was considered so urgent that the Director of Central Intelligence proposed that an investigation of UFOs be established on a priority basis throughout the intelligence and defense R and D communities, and urged an external research project of top-level scientists be established to analyze further the problem of UFOs. The proposal went nowhere.

Reports of UFOs, however, continue to pour into the National UFO Reporting Center[4] and other report collection agencies. Of the hundreds of thousands of sightings, there is general agreement that 5 percent are unexplainable in any other way than to accept them as extraterrestrial. This avalanche of irrefutable, undeniable evidence of UFO visits from beyond our solar system inundates the scientific community, month after month, year after year, but as of 2012 it has not converted even the least skeptical UFO-doubting physicist into a believer.

Everyone's reluctance to accept the reality of UFOs is perfectly understandable. After all, when you admit to the existence of UFOs, you have to accept as real their sudden appearances, accelerations, and decelerations and also their ability to hover at low altitudes, land, and suddenly take off and disappear. You have to take notice of the many different

4   www.ufocenter.com

vehicle shapes and sizes reported by observers. And by no means least of all, you have to accept that some humans probably have in fact been abducted, studied, transported, and returned to Earth.

Finally it dawns on you that these vehicles must harness an undiscovered law of gravitational space-time that voids what Isaac Newton taught us in 1687: the law of inertia. Control of inertia, as UFOs obviously do, is really a very useful tool. If inertia—the resistance an object offers to having its speed changed—can be controlled, it allows you to reduce to virtually nil the mass of anything you choose. For example, with inertial control one can effortlessly lift and move a stone weighing, let's say, three thousand pounds to a very precise location of one's choosing. A vehicle weighing nil can, with very low-grade energy, hover, almost instantly accelerate *every particle of matter contained in it* to hypervelocity, and disappear from sight in a fraction of a second. UFOs today show you this is possible.

Ultimately, people today have every reason to conclude superior—orders of magnitude superior—knowledge in fact exists, because our scientific community cannot otherwise explain the UFOs' extraordinary maneuvers. To put this in historical perspective, people today have no more idea of how flying saucers fly than did the people of the Aztec empire understand the sailing ships, armor-encased warriors, swords, horses, syphilis, and smallpox that Hernán Cortés brought to the New World in 1519. In that year many Aztecs were sure—*certain*—that what they saw wading ashore from Spanish galleons was "supernatural."[5] They were as mystified then by what they saw floating just off shore as people are today by the strange vehicles they see in the sky.

Most people scoff at the thought of visitors from outer space. Science and governments around the world label UFOs "supernatural" and unworthy of discussion. But with all this

---

5   Hugh Thomas, *Conquest: Montezuma, Cortés, and the Fall of Old Mexico* (New York, London: Simon & Schuster, 1993): 169, 170, 181, 186.

evidence available today, no thinking person should argue with a prediction that in the not-too-distant future incontrovertible proof of UFOs will appear.

CHAPTER 4

String Theory

Richard Feynman, an American physicist known for his work in expanding the understanding of quantum mechanics, once famously said, "I think I can safely say that nobody understands quantum mechanics." So I will not try to explain it to you and will only touch briefly on its main points.

Quantum mechanics, the physics of very small, very simple objects such as atoms, electrons, protons, and small particles of light called *photons*, describes much of the peculiar behavior of matter and energy at the subatomic level. Quantum mechanics was initially conceived as an explanation of why electrons remain spinning over time in their orbits, something explained by neither Newton's laws of motion nor Maxwell's laws of classical electromagnetism.

Physicists have folded theories describing three of the four fundamental forces of nature into quantum mechanics: electromagnetism, the strong nuclear force, and the weak nuclear force. But it has so far proven impossible to construct a quantum model of the elusive fourth fundamental force,[6] gravity, the basis of Einstein's equations in his theory of general relativity. This missing quantum gravity is an important issue in cosmology and is very disconcerting to physicists because it keeps them from formulating their long sought, elegant "theory of everything."

Einstein's general theory of relativity is the description of gravity used in physics, and, like quantum mechanics, its predictions had been verified in all observations and experiments to date. However, questions remain. Newton and Einstein state that the attractive strength of an object (its gravity) gets smaller as the mass of the object gets smaller, until an object's mass becomes zero and its gravitational attraction disappears. Unfortunately for Einstein's reputation, this change is true everywhere except for objects that are sucked into one of the thousands of black holes in outer space. There these objects get smaller in size and denser in mass as they fall farther into the abyss until they become what theoretical physicists call, for lack of a better word, a "singularity"—an imaginary particle with both zero diameter and infinite density. Einstein's general relativity predicts the existence of black holes but fails to account for objects in them that get smaller but denser, something physicists did not consider a problem until they eventually verified with indirect observations that black holes actually exist. Quantum mechanics doesn't address black holes. This means the laws of physics, as they are known today, have collapsed. It means theoretical physicists don't really understand anything about our existence. Black holes, a key to understanding the universe, baffle us.

In an effort to confront this dilemma, physicists in the mid-1980s began seriously to explore an extraordinary idea that had been kicked around for years. They replaced the dimensionless "particle of energy" in space described by conventional physics with a tiny one-dimensional "string" or

---

6   For a description of the four fundamental forces of nature, refer to appendix A.

loop, vibrating at certain frequencies the way violin strings vibrate at select frequencies to define their characteristics. They called their idea the *string theory* (later modified to *superstring theory*, and still later to *M-theory*). For simplicity's sake, we will call all the iterations of this theory: string theory.

Before string theory, mathematicians found that when they inserted infinitesimal numbers into Einstein's formulae, answers became infinity—in other words, absurd. Now, with string theory, Einstein's general relativity can, for example, explain those particles that are trillions and trillions times smaller than the diameter of a proton.

Many of the smartest mathematicians and physicists at the world's most prestigious institutions are working hard to perfect string theory. The tenacity with which they are attacking this problem should make everyone confident that experimental tests will soon verify string theory. When this finally happens, string theory will take its place beside general relativity and quantum mechanics to arrive at the long-sought "theory of everything," what scientists like to call the Grand Unified Theory (GUT).

Now, when one accepts "strings" as the basic building blocks of matter, as we one day will, it follows logically that one must accept string theory in its entirety, including all the special criteria that go along with it. One very special requirement of string theory is what string theorists like to call "superspace," i.e., space made up of eleven dimensions[7] rather than the three with which we are all familiar. So, when string theory is verified by scientific tests, eleven-dimensional space will supplant the three-dimensional space[8] you see around you.

---

7 String theorists have hypothesized over the years a variety of worlds, ranging from five to twenty-six or more dimensions. Most theorists seem to arrive at eleven dimensions, a number that will be used in this discussion to advance the argument.

8 Scientists like to think of our world as four-dimensional, time being the fourth dimension. Again, to simplify our discussion for the unscientific reader, I refer to the classic three dimensions.

If eleven-dimensional space seems absurd to the average reader, it is only the latest of the many "absurdities" proposed by physicists over the years. Einstein's general relativity calls for "curved space," and quantum theory counterintuitively predicts that small particles cannot be measured. It won't specify where an object is—only where it *probably* is. Objects can be in many places at the same time. Modern physicists argue for multiple parallel universes, some a distance no more than a millimeter from our own. Others, including Einstein in his day, argue back and forth over a paradox known as Schrödinger's Cat, a creature that is simultaneously dead *and* alive. Explanations and predictions about outer space and the infinitesimal have reached a point, as Feynman points out, where they are incomprehensible to *any* human.

Why can't man comprehend and visualize what modern physicists are teaching—concepts such as curved space, parallel universes, and an eleven-dimensional world? Quite simply, humans are *not equipped*, by virtue of the design limitations of their brains and five senses, to contemplate these concepts.

One of my contentions is that the counterintuitive explanations offered by modern physicists are clear evidence that, in their quest for a picture of fundamental reality, they have arrived at a "comprehension barrier." Physicists don't yet recognize that the explanations they are producing are counterintuitive for exactly the reason that the reality they are trying to describe is in fact an eleven-dimensional world, a world that only eleven-dimensional beings, with their superior intellects, are *equipped* to understand.

The difficulty people have accepting parallel universes, curved space, and an eleven-dimensional world is perfectly understandable. It can be compared to the difficulty a cocker spaniel has trying to understand calculus or even simple math. Our little puppy dogs love us and look up to us as gods. To them we must be gods. How else to explain our automobiles (which so many of them love to ride in), our homes, and how we find food for them every morning and night?

I predict that the Large Hadron Collider, the world's largest and highest-energy particle accelerator, constructed for the primary purpose of detecting the elusive Higgs boson, will also verify string theory and all its ramifications, one in particular being that we live in an eleven-dimensional universe. Sooner rather than later, eleven-dimensional reality will be accepted as fact by academia.

CHAPTER 5

INTELLIGENT DESIGN

Darwin declared in 1859 that all living things are descendants from a common life form, modified by unguided natural processes such as random variation, survival of the fittest, and natural selection. Darwinists claimed that, given enough time, a process called evolution explains the origin of entirely new species, organs, body plans, and even the spontaneous beginning of life on Earth.

Intelligent design, on the other hand, a theory developed by a community of scientists, philosophers, and scholars, uses proven laws of mathematical probability to conclude that some sort of "intelligent entity" designed our universe and the many species on Earth. Proponents contend the exquisitely formed components of living creatures, such as eyes, brains, and hearts, not only did not evolve by means of unguided, purposeless, natural forces of physics and chemistry but *could not have evolved* in the limited time that Earth has been in existence. Intelligent design answers many of the questions about the emergence of new and superior species and life forms that Darwin's evolutionists do not explain.

No one denies evolution is "change over time." Variation and natural selection do produce small changes within existing species, such as Darwin's finches and moths, clear examples of a process anthropologists call *horizontal evolution*, or *micro-evolution*. But Charles Darwin's theory is not just "horizontal evolution." It goes much further. It claims to explain how new, superior species evolve from inferior species, and even how living organisms arrived on Earth in the first place—a process called "vertical evolution."

For evolutionists, vertical evolution is an assumption—nothing more than a guess, in other words, a hypothesis. It is, in fact, one for which there is no hard evidence. There is a "missing link" between one species and its immediate superior species no evolutionist can explain, so one is forced to accept vertical evolution on faith. No one has ever found evidence of or observed the origin of a new species by variation and selection, much less the origin of new organs and body parts. Not even modern genetics has solved the problem. Geneticists often use fruit flies in their experiments, because of their short lifespan, and they always find that no matter what they do to the DNA of a fruit fly embryo, only three possible outcomes exist: a normal fruit fly, a defective fruit fly, or a dead fruit fly.

Opponents of intelligent design are horrified by the "intelligent being" aspect of the argument, and they vigorously defend evolution because science today requires that everything be explicable by natural causes. Science is limited to what can be seen, touched, and, most importantly, tested. Intelligent design assumes a mechanism existing outside of nature, which we cannot see, cannot replicate, cannot control, and thus cannot test or subject to any natural examination. If unnatural explanation is used, anything can be explained away, because it's not a testable hypothesis.

But one is particularly taken by the harsh, vitriolic—almost apoplectic—nature of the arguments and attacks against intelligent design. Can it be this reflects acknowledged but unspoken existential flaws in Darwinian theory?

One of the few examples of restrained argument against intelligent design occurred in a court case when Darwin's theory was not actually in jeopardy. The question was whether intelligent design was or was not "religion." In 2005 Judge John E. Jones III ruled, in US District Court for the Middle District of Pennsylvania, that intelligent design was creationism—"religion," in other words—and was therefore forbidden by the US Constitution in public high schools. The school board involved in the case was forced to pay exorbitant court and legal fees, thus discouraging other schools from even considering the inclusion of intelligent design in their curricula.[9]

Many of our finest scientists rebel at the concept of a superior intelligence designing our universe. It invokes the supernatural, an untestable hypothesis that when applied to any occurrence automatically ends further inquiry. Ever since Newton's day, scientists have limited their explanations of physical occurrences to natural physical laws. To many observers, however, the efforts of scientists to avoid consideration of intelligent design seem, ironically, the reverse of the refusal of fifteenth- and sixteenth-century theologians to consider that natural laws caused Earth to circle the sun.

Many see Darwinists' arguments as weak, akin to Simon Newcomb's very plausible prediction that human flight was impossible—made just three months before the Wright brothers flew their airplane at Kitty Hawk. Fallacious Darwinist arguments are everywhere. Evolutionists argue, for example, that apes and humans have nearly identical combinations of genes, proving, they think, one must have evolved from the other. Intelligent design advocates reply that of course many species have mostly identical genes, just as automobiles

---

9   Tammy Kitzmiller, et al. v. Dover Area School District, et al., 400 F. Supp. 2d 707, Docket no. 4cv2688.

have mostly identical parts—wheels, fenders, windshields, and tailpipes. Do Darwinists therefore believe Chevrolet automobiles *evolved* from Fords?

Intelligent design theorists have excellent points to offer. Fact-based rebuttals by Darwinists, rather than courtroom theatrics, rebukes, and referrals to the first amendment of the United States Constitution, would refute them if they could be made.

Not all scientists searching for ancient man are rigid in their thinking, of course. In a 1990 PBS documentary, famed paleoanthropologist Richard Leakey stated unequivocally that he believed nothing had been found up to that time to suggest a gradual evolution of the humanoid species, including his "Lucie" fossil discovery, and that there was more evidence indicating an abrupt arrival of man. His wife, Mary Leakey, a noted anthropologist in her own right, said something similar just before her death, in 1996, at the age of eighty-three. Ms. Leakey was convinced that man had evolved from ape-like ancestors, but she was also sure scientists would never be able to prove a particular scenario of human evolution. Three months before her death, she said in an interview, "All these trees of life with their branches of our ancestors, that's a lot of nonsense."

Compared to the science of evolution, the science of *abiogenesis*—the study of processes by which biological life spontaneously begins from inorganic matter—is still in its infancy and at something of a dead end. This is the other huge hole in Darwinian theory. Biologists cannot explain life's emergence from a prebiotic world four billion years ago—a primordial world some call "rock soup."

If there is any doubt evolutionary theory is in turmoil, look up the comments of journalist Mark Davis regarding his research for a story on Neanderthals for a NOVA program first broadcast January 22, 2002. He concluded that everyone he talked to thought the last person he interviewed was "an idiot." He found during his interviews that scholars called each other everything from "politically correct" to racist. The more anthropologists he interviewed, the more confused he became.[10]

Mark Davis is not alone in his confusion, even if he didn't see the reason for his confusion. He did not recognize how, in their efforts to avoid any suggestion of a supernatural force at work, paleoanthropologists tie all their discoveries to Darwinian evolution and, thus, to a "tree of life," with branches and a trunk leading down to that very first seed of a living organism that formed in the oceans. Every time a new fossil is discovered, there are new arguments over where it should fit in the list of species, and, often as not, names are invented or revised to allow the new discovery to fit with the others.

Each fossil category, or *taxon*, is vaguely defined and is often adjusted to allow for each newly discovered bone. What, for example, defines, anatomically, a member of the genus *Homo*? The genus *Australopithicus*? How about the genus *Ardipithecus*? How do you tell the difference between a *Homo sapiens* and a *Homo sapiens sapiens*, a term used by some taxonomists to differentiate modern man from less modern man? Or is there a difference? To a new student of anthropology, the taxonomy of Paleolithic bipedal creatures appears to be in shambles.

Instead of trying to trace every fossil find back to evolution, paleoanthropologists could more clearly categorize their fossils if they used a variation of the Dewey Decimal System or something like the American military's system for categorizing its World War II airplanes—numbers and letters to categorize age, brain capability, teeth, heads, arms, legs, pelvises, etc. Fossil age, sizes, shapes, and brain capacities would instantly organize into a comprehensible picture. Not only would the picture make sense, but a place would be found for

---

10   Refer to www.pbs.org/wgbh/nova/neanderthals for an interesting discussion of Davis' investigation into this contentious field of study.

all subsequent discoveries in the file without raging controversy and rearrangements of the system.

What should be obvious to paleoanthropologists, and what had been staring them in the face for at least fifty years—what some would say is "the elephant in the room"—is that they become hopelessly entangled in controversy whenever they try to force-fit every new fossil they find into a branch of the tree of Darwinian evolution. Sooner or later they will recognize that each step of progress the humanoid creature has made over the eons, from his frolic in the trees of Africa to his obsession with the iPhone, is nothing more or less than the result of successively smarter, more complex brains provided to the creature, along with appropriately improved body structure by what can only be an intelligence of some sort. Not undirected chance.

* * * *

I have made an objective survey of the arguments about evolution and abiogenesis, both with pros and cons. It brings me to conclude that anthropological taxonomists will eventually realize categorizing bipedal fossils is no different from categorizing the fossilized remains of various parts of automobiles designed and built from 1890 to 2000 that may someday be discovered under layers of earth, let's say a million years from now, by some far-advanced civilization. Taxonomists will conclude that intelligent designers employed in the many automobile companies that came and went during that period were the ones who experimented and "evolved" the blueprints—the DNA, if you will—of those cars.

UFOs and the eleven-dimensional reality of string theory will be verified, and when they are, the key argument against intelligent design—reliance on the supernatural—will vanish, and physicists will be free to conclude that their universe was designed by an intelligent entity. Darwinism will drift away from serious consideration. Evolutionary theory will evaporate just as the theory of earth-centricity did four hundred years ago, leaving mankind with a supernatural explanation for life's existence on Earth, a thought quite repugnant to most scientists today.

A key lesson will be that the intelligent designers who designed mankind, the planets, galaxies, and the entire universe, cannot possibly be intellectually advanced three-dimensional aliens from another planet within the universe, because three-dimensionals (humans) can't even grasp such relatively simple concepts (for intelligent designers) as curved space, quantum mechanics, and parallel universes. To have worked with and shaped these humanly incomprehensible concepts to design the universe, intelligent designers necessarily must have powers far beyond three-dimensional human intellectual capability.

* * * *

So now you have my prediction: that UFOs, string theory, and intelligent design will be scientifically verified in the near future. For people today it is a lot to digest. Many of you certainly still have doubts—this is perfectly understandable. But recall that it took generations for a majority of scientists to accept Isaac Newton's theories and assumptions. For heaven's sake, it took decades for medical doctors to accept Louis Pasteur's germ theory.

If and when, however, you accept my predictions and their eventual scientific confirmation, the world beyond today, a world of entirely new certainties, opens before you. It is a world—the real nature of reality—that I will describe to you using nothing more than simple deduction and logic.

CHAPTER 6

Thoughts

When you accept my predictions for the world of tomorrow and beyond—that is, when you accept that UFOs are vehicles constructed of a hyper-advanced technology every bit as incomprehensible to us as were Cortés's galleons, swords, and armor plate to the Aztecs; when you accept that the theory of intelligent design trumps the fading theory of evolution; and when you know physicists will confirm string theory and the eleven-dimensional other-world that it requires—you have entered a world comfortable with new horizons and concepts, new "certainties," if you will. You will have mastered the last three of the many clues needed to unravel the mystery of true reality.

What our various religions called "God" is now our "Intelligent Designer," or "Maker," an entity—not just an individual—capable of designing our entire universe and every plant and animal, every microbe, in it. You know intelligent designers established the physical laws, physical constants, and rules that specify how everything in our universe works—all the rules and physical laws discovered and reported on by Michael Faraday, James Clerk Maxwell, Louis Pasteur, Albert Einstein, and the other scientists who followed a path first illuminated by Isaac Newton. After all, this is what an intelligent designer is. It is what an intelligent designer does. Our "intelligent designer" is no longer considered "supernatural" by the scientific community, but, rather, "extra-terrestrial." Maybe the term "extra-galactic" is appropriate.

With the acceptance of eleven-dimensional reality, intelligent design of our universe, and vehicles beyond your comprehension entering and exiting your solar system, it is clear that our intelligent designers are eleven-dimensional beings from an eleven-dimensional world. Your meager five senses and limited brain capacity are entirely unequipped to envision or understand any of this.

With this understanding, you appreciate that designing a universe such as ours with all its complexities, intricacies, and life forms requires a huge design bureaucracy—a Committee—and that universe design is no more difficult for intelligent designers who know how to do it than designing a modern jet-liner is for engineers at the Boeing Aircraft Company. You understand the designs of Earth's living plants and animals are as carefully crafted as are the jet airliners of today. And just as employees of Boeing cherish their aircraft, just as employees of Apple are proud of their iPhones, and just as employees of the Ford Motor Company love their Mustang automobiles, you know that each intelligent designer loves and cherishes in exactly the same way the world he helped to create, and that he hopes for the very best possible outcome for each and every living plant and animal in his experimental world.

Building on my three assumptions, or "axioms" as I like to call them, you can deduce how it all began. One can hypothesize that a Committee of intelligent designers initiated a project to model a primitive (by their standards) civilization. Their reason for doing so is still, of course, unknown, but go ahead and speculate. Perhaps the reason for the project was simply to demonstrate how a universe can be designed, constructed, and observed as it progresses on its own from non-life to life and onward to civilized existence. Or perhaps the object was to demonstrate that the Committee members could design a creature that would civilize itself and not self-destruct. Maybe they were looking for clues and evidence to explain their own origin.

In any event, the universe these universe designers fashioned mimicked their own eleven-dimensional universe, but on a simplified scale. What could be simpler (by their standards) than a three-dimensional (height, width, length) universe? (Recall that Walt Disney drew his characters with four fingers rather than five, to simplify and speed up their creation.)

A Committee of designers, engineers, administrators, clerks, technicians, and advisors was formed to design and manufacture a universe populated by living plants and animals. For simplicity, all the living organisms in this world were designed around a single molecule with the flexibility to produce life in almost infinite variety. This single, ultra-complex molecule was the key to and the genius behind the project's success. (Hominid scientists discovered this molecule some years back and named it *deoxyribonucleic acid*: DNA.)

The trick was to use simple compounds, such as sugars and phosphates, to produce a double-helix-style molecule that can function in a three-dimensional world. This molecule had to be orders of magnitude less complex than the quadruple-helix molecule that is at the core of the designers' own existence. At the same time, engineers had to perfect computerized laboratory equipment that could program or mutate this exquisite molecule in all the subtle ways necessary to produce the countless living organisms, both plant and animal, needed for the project.

One can only imagine the trials and errors, engineering committee meetings, arguments, intellectual conflicts, and stress and exhaustion that went into getting a single molecule, acceptable to designers, consultants, midlevel bureaucrats, and finally top management, approved. Everyone knew this molecule had to perform to near perfection in order for the grand project to succeed. All the sundry living organisms this molecule produced had to thrive, self-repair, find their own food, and, most importantly, replicate. A tall order indeed.

With this very special building-block molecule and its programming apparatus finalized, project managers needed to agree on a general body plan for the living organisms that would populate their new world. After more long meetings, let's suppose designers decided that plants would extend into the surface of Earth as well as upward into the atmosphere for their sustenance and support. A major design decision divided the animal world into those with and those without backbones (vertebrates and invertebrates). Invertebrates were freeform in design, meaning the designers could create without limits. Vertebrate construction required extra thought, and finally engineers derived a universal body plan consisting of a main body, four limbs, and a head, that housed a computer controller. Each living life form had sensors appropriate and sufficient to allow it to react to and survive in its own particular environment.

When our Committee completed its designs, collected all the supplies and materials needed for its experimental universe, and set aside laboratory space for its universe of galaxies, stars, planets, and black holes, it initiated a controlled explosion—a "big bang"—to begin the experiment. Their experimental universe expanded and eventually cooled sufficiently for life to exist, and technicians searched for a location (planet) within this universe suitable for their needs. Earth, with its benevolent environment, was an obvious choice, perhaps one they had singled out before the experiment began. It's possible but unlikely that other planets in their universe had all of Earth's precisely calibrated characteristics[11] to sustain their experiments. But why complicate the experiment with duplicate efforts?

At first, creatures and plant life introduced to the planet were simple, single-celled organisms living in the oceans. Later, as encouraging results came in, engineers designed and placed more complex, multi-celled plants and creatures in the seas. When the designers saw these creatures flourish, they took the next step and carefully introduced simple life forms onto dry land. The less successful designs faltered and became extinct, but as some of the more sophisticated DNA creations succeeded, designers built on their success and produced creature and plant-life designs of still more complexity. Before long the Committee had designed complex four-legged vertebrate animals capable of running at high speeds, some huge in size, others very small, all coexisting in a "kill or be killed" society.

Eventually Committee designers found progress and further complexity pretty much impossible. Progress stalled, until at some point it occurred to one of the bright Committee engineers that if their creatures were ever to exhibit more complexity, there would need to be a huge leap in design technology. Working nights and weekends, this engineer, with the help of a lab technician, mutated a DNA molecule so that a creature's pelvic area and legs allowed it to stand upright, and converted its front legs into arms with five-digit appendages and opposable thumbs capable of carrying goods and producing tools and weapons. And, of paramount importance, our clever engineer designed for the brute a new, higher-capacity brain, encapsulated in an enlarged cranium.

Here at last was a creature construct, a sophisticated machine, if you will, with a complex brain that formed and communicated ideas to others, produced goods, and adapted to the changing environment around it. This was a huge step

---

11   Refer to Appendix B for a discussion of some of these characteristics.

forward—the very paradigm change the Committee had hoped for—both in the design of the body framework and the brain that controlled it. The project was immediately seen as so huge that our clever engineer was directed to turn over development of his new creature to the many other engineers and designers in the bureaucracy who could specialize in and devote their talents to detail problems. Sprinkled among them were less talented engineers with outsized egos that had been bruised by the praise heaped on our clever engineer and who were eager to promote their own agendas wherever they could.

Numerous design variations, or "models" as they call them, were proposed and tested in their laboratories. Some advanced so far as to be placed on Earth for field tests. Over time, many experimental models[12] were introduced onto the planet for trial to determine if these bipedal, ultra-complex creatures could survive and, hopefully, thrive. Some survived for hundreds of earth-years, some for thousands of earth-years, some for millions of earth-years.

Once the experiment began, it was of course necessary to observe results. What better way to observe than by way of the vehicles you have heard about for generations: UFOs. Immune to Newton's first law of motion—inertia—that inhabits all earthly matter, UFOs travel at speeds that vastly exceed the speed of light. With such vehicles observers quickly transited their entire prototype universe and reported observations and data back to their home base. When the vehicle was slowed to slightly less than the speed of light, its passengers manipulated, by means of a phenomenon known as *time dilation*, the point in Earth's history at which they arrived on the planet.[13]

---

12   See Appendix C for a complete listing of the bipedal species models.

13   Einstein was the first to recognize time dilation, i.e., that travel near the speed of light significantly distorts the traveler's time relative to Earth time. Today the time dilation effect is confirmed by calculations and numerous examples: the GPS satellite navigation system and the clarity of TV picture screens depend on time dilation, to name just two. Visitors arriving in Earth's past are prevented, of course, from affecting change in any way. If they visit the past, it is only to observe past events.

But only so much can be learned by looking down from the sky. For close observation, it seems likely—certain really—that the Committee developed a technique by which three-dimensional observers, or "earth-monitors," as they called them, were introduced onto the surface of their experimental three-dimensional world. For these special on-site earth-monitors to be effective, they had to closely resemble the experimental humanoid creatures populating Earth but also had to have superior personal qualities, to enhance their chances for survival in this experimental world. To give these earth-monitors historical perspective, perhaps they lived far longer than Earth's inhabitants. Linear, serial time, Newtonian laws of motion, and Einsteinian speed-limit constraints imposed on you did not bind our visitors.

All experiments need detailed feedback, of course. To this end, each hominid was given a distinctive identity, a "soul" that carries its host's personal characteristics and collects its life's memories, emotions, and thoughts. Constructed in eleven-dimensional code, and thus undetectable to its three-dimensional host, the soul, after the demise of its earthbound life form, returns to a Committee database for review, evaluation, and final judgment. Rest assured, this database is more than just a repository for collecting engineering data. It is a place where the souls that pass judgment are welcomed home, to a place where they can reunite with souls they knew and loved on Earth and receive the gratitude of their eleven-dimensional superiors. Souls found to be tinged with true evil in any of its several forms are immediately diverted to a special laboratory in the DNA design office for dissection and study, a most unpleasant but very important process.

It can be estimated that 108 billion[14] such souls have returned to the eleven-dimensional database since the beginning of the program. This database is crucial for providing

---

14   The Population Reference Bureau estimates 107, 602,707,791 humans have lived on Earth (October 2011).

detailed information to the Committee, so it can improve its bipedal creature design and bring out new, improved models.

With these systems of observation and collection in place, a huge variety of plant and creature models was introduced to the planet over the earth-centuries and left to prosper or drift into extinction as other, newer, updated models come along. How similar is this process to what auto companies do on a regular basis, bringing in new, more modern automobile models? Early hominid models, such as Australopithecus *afarensis* (their Model 5), *Homo habilis* (Model 7), *Homo neanderthalensis* (Model 19), and Cro-Magnon man (Model 23), all clearly at dead ends and unable to progress past early stages of their development, were allowed to pass away or be exterminated by follow-on designs.

Finally a new humanoid model, their twenty-fourth fundamentally new model, *Homo sapiens sapiens*[15]—named by none other than the model itself—was placed on Earth at many different locations and was found, after a slow start, to thrive. At long last an experimental Committee-designed creature—known to them as M24 (for "Model 24")—exhibited the "behavioral modernity" they were looking for, i.e., behaviors such as language, laughter, religion, art, music, cooking, games, and jokes, which separate modern man from its predecessors.

Homo sapiens sapiens progressed beyond the wildest expectations of the Committee, given its many failures up to that time. This "modern man" quickly moved from the "hunter-gatherer" stage, in which all its predecessors had been confined, to an agricultural existence, at once vastly reducing the land area required for an individual's sustenance, allowing it to remain in one location, where buildings and

---

15   Many if not most paleoanthropologists add a second "sapiens" to the classification of fossils of "modern man." Everything I can find on the subject indicates the modern man we see all around us did not appear until 11,500 +/- 3,000 years ago. None of the creatures existing before that time exhibited the characteristics and behaviors necessary to achieve what modern man has achieved in less than 15,000 years.

other infrastructure could be established. Of monumental importance to designers was that, for the first time, their new creature discovered and exploited copper, tin, iron, oil, and all the ninety-some elements residing for their use under Earth's surface. At last the Committee's new creature design multiplied hugely in number and, in doing so, established territorial boundaries and governmental institutions.

But logic tells you all this incredible progress must have come with a heavy price. Reports by earth-monitors showed the creature's DNA falling short of design specifications in key areas. Moreover, certain of the humanoid characteristics, useful while *Homo sapiens sapiens* emerged from its primitive existence, are now, in a modern world, deficiencies that endanger the project. The Committee sees that outright design errors got past inspectors. The most egregious: a predisposition for evil in about 10 percent of the creatures. These are serious impediments to further progress. By now their smart new creature has invented nuclear, biological, and chemical weapons capable of rendering the planet untenable for *any* living organism, thus ending the experiment if they are used in quantity. And some of these weapons are falling into the hands of gravely deficient hominids.

You do not have to be reminded that in the design of complex products, bureaucratic and engineering mistakes inevitably occur. In the case of jet airliners, for example, serious errors are usually revealed after a tragic plane crash. When aircraft design flaws are discovered in the course of a safety committee's accident investigation, design upgrades are made.

Until the advent of supercomputers, it was whispered in the aircraft industry that an airliner wasn't fully proven until it had killed five hundred passengers. Perhaps it can be said a new hominid design isn't fully proven until it kills a half billion of its own in wars, social turmoil, and illness. Sooner or later the Committee will conclude that design errors are the cause of their hominid's misadventures.

Logic tells us that somewhere a Committee of eleven-dimensional intelligent designers is working to correct the design flaws in the hominids they placed in their universe. And it is logical to expect that when these design flaws are corrected, the Committee will report its findings and recommendations to their earth-monitors for their final assessment and comments before they introduce to the planet a new creature design, a new and superior hominid model.

* * * *

The foregoing scenario was the product of nothing more than pure logical deduction and took you well beyond my three predictions. Its pieces fit together perfectly, like a hand in a glove, like the pieces of a jigsaw puzzle. You could have produced other scenarios from these three predictions, of course. But, as we will see, they would have been wrong.

CHAPTER 7

A Scenario for 2030

May 9, 2030, is a beautiful spring day in the small New Mexican town of Nogal; temperature about sixty-five, with clear blue skies. I am visiting Harry Bates, a fellow earth-monitor, who is at that advanced age when recollections of one's past seem to take up more and more of one's waking day. Much has happened in the last thirty years to stir the very foundation of human beliefs.

At first it was the gradual acceptance of intelligent design in academic circles and the subsequent inclusion of these ideas in high school curricula. The Darwinian evolutionists had fought bitterly against what they saw as an atrocity to scientific reason, but as physicists announced more and more progress in their search for "strings" and eleven-dimensional reality, the strongest Darwinian argument against intelligent design—that it was "supernatural"—receded. The longer the aberrant creationist argument persisted and grew, the more courts, school boards, and politicians alike began to waver. They finally give in, one by one, to at least broaching the subject of intelligent design in the schools.

The world's religions watched closely as these ideas gained acceptance over time, and eventually most religious teachers grew to accept them and modify their teachings to accommodate the new ideas. After all, none of the recognized religions had ever before attempted to define exactly who or what "God" actually was. (Islam was another story, of course. The mullahs, imams, and sheiks united in complete denial, and as Harry saw it, were to this day "bouncing off the walls of their mosques" in defiance.)

The turmoil over the acceptance of intelligent design had not yet ended when physicists at the Hadron Particle Collider near Geneva, Switzerland, announced proof that the long-debated string theory had at last been shown to bring Albert Einstein's theory of general relativity into complete agreement with quantum theory. At long last, a "theory of everything," something Einstein had labored over his entire life, had been developed and accepted by a majority of the world's physicists. All of the ramifications and fallout from this revelation were still not fully accepted by the public, but the theory's requirement of eleven-dimensional reality was being increasingly pressed home to world populations by an eager scientific-academic media.

And just two years ago, an encounter with a vehicle from outer space had at last proven aliens' existence and immense intellectual power. An extraterrestrial vehicle had landed in a field near Centreville, Virginia, near the Civil War battlefield of Bull Run, and from it emerged a group of figures clothed in tight-fitting white garments, motioning to anyone attempting to approach their vehicle to keep his distance.

Even the most strident doubter now could not refute the evidence. A media that had once scoffed at "flying saucers" now compared the UFO doubters to the silly individuals who had scoffed at the possibility of a "flying machine" prior to the Wright brothers' flight in December, 1903. To movie buffs the encounter brought back memories of the 1951 movie *The Day the Earth Stood Still.*

The vehicle, surrounded by a dozen or more smaller craft, had landed in a space the size of a football field and was so huge the rounded "points" of its triangular shape extended into neighboring pastures. It had arrived so suddenly that military assets could not respond until after it had landed, and then they were kept at a distance by smaller craft circling about, shepherding the fighter jets and ground-based military units attempting to intervene, away from the grounded ship. It was an incredible sight, one well documented by the world's media.

When the crowds and the military units surrounding the vehicle had finally stabilized and calmed sufficiently, an opening appeared in the bottom of the craft and several figures appeared. They approached one edge of the crowd and motioned to several bystanders that it would be safe for them to come forward to meet the leader of the group. To the humans' amazement, the group leader introduced himself in clear English and asked that an official representative of their nation be brought to him for a discussion of his visit to Earth.

In the days that followed, these figures explained their vehicle had experienced mechanical difficulties and would be departing as soon as repairs were made. They further explained that they represented an extraterrestrial entity interested only in world peace for Earth's inhabitants. Their explanation was that, for them, Earth was a primitive society of no interest other than as a curiosity worthy of their observation. The visitors added that their tight-fitting suits, which totally enclosed them, were for their protection against earthly microorganisms, for which they had no antibody protection. For this reason, no one from Earth was allowed to enter their vehicle.

After five days the craft was repaired, and they departed as suddenly as they had arrived, leaving huge questions for politicians and the talking heads on TV to discuss ad nauseam.

The unintended visit of the extraterrestrials convulsed the world's population. Almost daily now new religions were invented out of convoluted interpretations of the visit and preached by neo-evangelists to a gullible public. Politicians outdid each other with flowery language, appealing for calm and begging for unity. Ominously, several nuclear-armed countries controlled by suicidal religious lunatics were moving to ignite war, precisely what the recent visitors to our world had counseled against. For a Committee that had been so tolerant of their spectacularly successful but sadly flawed M24 for so long, it was the last straw. They could not ignore the prospect of nuclear Armageddon in their experimental world. Something had to be done.

They turned to the one member of the Committee who had not only championed the Earth project from the beginning but who had actually undergone the rigorous process of transformation into a being capable of visiting and living in the Earth environment for extended periods. Let's say his name is VAATU-9. He was tasked with the job of eliminating flaws, especially evil, from the M24, even if it meant bringing forward a new species model.

VAATU-9 worked quickly and soon completed a report comprising a review of the M24's past history, its flaws, and its prospects for the future. When the report was complete, he called a meeting of the Committee and presented his conclusions to them: the M24 must go. Its place would be taken by the Model 25 (M25), their newest hominid model, one that had been in development for years. The Committee accepted his recommendations and ordered VAATU-9 to present his report to the earth-monitors at the earliest moment and to hear firsthand their reactions and recommendation for the best means to expunge the M24 from the planet.

Inexplicably, a copy of VAATU-9's report fell into the hands of a New York City newspaper reporter.

CHAPTER 8

The Visitor's Report

Elliot Flindecker III, a short, pudgy, bald-headed man with round glasses, rounded shoulders, and large ears that seem to press against his scull, is a typical small-town newspaper editor with a flair for producing the written word. He edits the only local newspaper to which Harry Bates has access.

This particular morning, Mr. Flindecker featured a story he had received from the Associated Press about an "alien's" report. It was so lengthy Flindecker had to summarize much of it, and it was so explosive he began with a detailed explanation of it in his own words.

The newspaper article and accompanying report, which I insert here in its entirety, supports and expands on the deductions I presented to you in chapter 6. The article began with this introduction, and, other than advertisements, filled nearly the entire newspaper that day.

REPORT DETAILS EXTINCTION OF EARTH'S HUMAN POPULATION
HUMANS TO BE REPLACED WITH SUPERIOR BEINGS

AP, Washington D.C., July 17: US government officials today announced discovery of a report written by an alien, titled *The Visitor's Report*, detailing a plan to replace the human race with superior human beings. According to reports, paragraph after paragraph of the lengthy, detailed report looked like this:

The NSA had little trouble deciphering the document. In so doing, it replaced descriptions and terms unique to an alien world with terms common to our readers.

According to the report, an individual named VAATU-9 wrote it. He and his community, or research facility, as it is described in this report, exist in an eleven-dimensional world comprised of life forms alien and superior in all respects to anything seen on Earth. Normally invisible to those of us on Earth, these alien life forms, according to this report, visit and move about our world at will in their vehicles, sometimes at incomprehensible speeds. They can, if they so choose, make themselves and their vehicles visible to us, their experimental subjects, and when they do, they usually, but not always, transform their appearance, chameleon-like, to conform to their current surroundings, a process known to scientists as *hetromorphosis*. Many of those who are transformed blend into our society, become "earth-monitors," and report their observations back to their superiors.

Other than the routine of daily life, earth-monitors make no effort to affect Earth's world events or the lives of people. However, earth–monitors do hear our wishes, including our prayers, and relay messages via their vehicles overhead, to their distant facility by gathering twigs, grasses, and leaves into a small pile in front of them, and, kneeling down as if to pray, they mutter incoherent words, phrases, and grunts that only their superiors can comprehend. On rare occasions their parent research facility will alter earthly activities or improve individual circumstances in response to these pleas.

The Visitor's Report explains that a research facility experimentally produced our universe and tells how it was done. When design work was complete, an area was set aside in its laboratory for the experiment. Materials were obtained, measured into precise quantities, and compacted into micro-particle size. When all was ready, the mixture was caused to explode. After the resulting mixture cooled sufficiently, inspectors were sent off into the experimental space to search for a location that perfectly matched forty-one exquisitely detailed characteristics they had designated as requirements for three-dimensional life. These requirements were so daunting, and the odds against finding such a location so huge, that multiple creation attempts, or "Big Bangs," as scientists like to call them, were required before inspectors could find a suitable location for their life forms.

The fourth "big-bang" attempt was successful. It then took five billion earth-years to allow the mix to cool and coalesce into basic elements, stars, and planets. Over the next several million earth-years, searches revealed a planet with all the necessary life-supporting characteristics. When its molten state, and later its miles-thick ice covering, evolved into quiet oceans and dry land, single-cell life and food for its sustenance were introduced to the water. As these simple creatures proliferated, multitudes of multi-cell creatures and plants, created with DNA molecules modified by fractal-generating software (distantly related to what was popularized in the 1980s by Benoit

Mandlebrot and many others) were introduced first to oceans, then cautiously to the land.

Finally, after thirteen billion earth-years of site preparation, it was deemed safe to deploy, to various suitable locations on planet Earth, the first of the experimental humanoid bipedal creatures they had created: Model 1 (or M1), classified by Earth's paleoanthropologists as *Sahelanthropus tchadensis*.[16] Their objective with M1 was modest: to look for and evaluate design problems associated with creature bipedalism. Each successive creature created in the early design series, Models 1 through 16, were fitted with increasingly larger, more capable brains, housed in cranial vaults increasing in size over time from seven hundred to thirteen hundred cubic centimeters. As brain capability increased, the creatures eventually tamed fire and learned to butcher the larger animals and to adjust somewhat to changing weather.

But none of them progressed beyond a level of achievement exemplified by the Acheulien stone hand ax. The Model 17 design, *Homo erectus*, brought forward around 1.6 million earth-years ago and possessing an improved larynx to enhance the creature's oral communication, was able to produce several additional useful small stone tools. But there its progress plateaued. The hominid design project then languished until taken up again by a new research facility administration that saw a way to significantly improve the concept.

Three hundred fifty thousand earth-years ago, experiments began on a new bipedal creature design, the Model 19 series, later named *Homo neanderthalensis* by Earth's paleoanthropologists. This new creature, equipped with a cranial vault of 1,700

---

16   Refer to Appendix C and D for a description of all the bipedal creature models.

cc., was larger and more robust in stature, indeed an improvement over *Homo erectus*. It produced better stone tools, could better adapt to cruel weather conditions, and exhibited faint traces of cultural creativity. However, over some three hundred thousand earth-years of existence, it could do little else to improve its environment.

By 50,000 BC, researchers had come to realize that if Earth hominids were ever to demonstrate an ability to civilize themselves, an even newer, upgraded design, classified Model 23 and named *Cro-Magnon* by Earth's paleoanthropologists, would have to be designed and deployed. Frustratingly, within thirty thousand years of its introduction, it was again distressingly obvious this new creature, still unaware of the treasures placed in the earth's crust for its use—oil, coal, metals, minerals—was, like its predecessors, not going to progress past the stone tool.

Undeterred by past failures, work feverishly began on a still newer bipedal creature. This time the engineers had available to them new, vastly improved laboratory test devices and techniques that could weed out creature weaknesses at the DNA level. By 11,500 BC, just as one of Earth's many glacial cycles was ending, the research facility had progressed with their creature upgrade, designated Model 24 (M24), as far as laboratory trials and evaluations could take them, and yet design questions persisted.

In a hurry to take advantage of receding ice and a warming Earth weather cycle, before the M24 was fully evaluated and tested, designers deployed their new creatures to several dozen widely dispersed locations along rivers in warm or mild environments. Placed in breeding communities ranging in population from fifty to two hundred, divided evenly between male and female, all individuals had identical DNA, and, therefore, equal capabilities. They differed only in distinct facial and skin features, which enabled researchers to identify their original geographic placements. These identifiers allowed the Committee to make allowances for the effect location had on results. Earth's taxonomists identify these variants as "race."

By now the Committee realized these early humanoid bipeds, with their gracile physical architecture could not survive in their primitive surroundings without tutoring. How could they ever imagine where to look for food and shelter without help? It was decided instructors would be on hand at the start-up of each of the original communities to instruct them how to survive in their new habitat. After each new colony established a foothold, the start-up instructors that they had come to know and love backed off and replacements, called earth-monitors, blended in, continuing unobtrusively and covertly to observe outcomes.

All creatures are endowed with preprogrammed survival instincts, but with the more advanced designs, such as the hominids, the instructors at the scenes of the implantations provided extensive training and practice on survival techniques. This protocol included where to look and how to recognize nourishment when they see it. Instructors taught them rudimentary language and primitive techniques on how to prepare clothing and shelter.

Instructors stayed with these advanced designs for years until they were established to their satisfaction in their new surroundings and they could be replaced by passive earth-monitors. In doing so, later generations, hearing stories handed down by word of mouth, evolved a belief in extraterrestrial beings, or 'gods' as they called them. These stories became embedded into their belief systems. To strengthen these beliefs, these early instructors helped them construct their first temples, churches, and tombs—some, such as the pyramids, quite pretentious.[17] All early religions believed in multiple gods, just as there were multiple monitors guiding the original implantees in their colonies.

At long last, around 8000 BC, in Turkey near the Syrian border, at a place called Gobekli Tepe, earth-monitors reported what the research facility had long been hoping for: evidence of fully modern language skills, the capacity for abstract thought, and symbolism to express cultural creativity. As time progressed, and in numerous other locations, the M24 invented complex spoken languages expressing ideas, humor, and emotions with reasonable precision. For the first time ever, their experimental creature produced art, music, games, and jokes in abundance, and particularly enjoyed inventing and telling stories to each other.

The new hominid looked to the sky, noticed the stars in the heavens, and speculated about supreme beings or gods, patterning them after hazy and very distant memories of the instructors who had guided their founding generations. The end of "hunter-gathering" of food and other essentials had arrived. Research facility administration officials were thrilled to find the M24 was, finally, on the road to self-civilization.

Their new creature cultivated crops and harvested them for food, allowing for significant increases in

---

17 On a recent visit to Egypt, your author was surprised to hear that out of all the hieroglyphics, papyrus records, and artifacts depicting its vast history, not a single record exists describing how their pyramids were constructed, not a one. Regarding the two principal hypotheses suggesting (guessing, actually) how the pyramids might have been built—large ramps or mechanical levers—arguments against them are more persuasive than those agreeing with them. How they were erected remains a puzzle.

their populations. Above all, the new creature, the M24, discovered and exploited minerals that had lain under Earth's surface for its use. With a huge sigh of relief, Committee designers realized they had finally learned to program their basic DNA molecule to form a brain with the power to discover and exploit the prodigious quantities of minerals, energy sources, and materials they had placed for its use on and under the planet's surface—riches unrecognized for millennia by all previous humanoid models. Furthermore, their newest humanoid model had the brainpower to manufacture these treasures into useful and very complex goods. *Homo sapiens sapiens* learned to smelt tin, lead, copper, and iron for its benefit, eventually learning to manufacture from these treasures almost anything it desired, from efficient, sometimes huge, vehicles for travel on land, sea, and air, to very small and sophisticated communication devices.

The project was a success—or was it? To the Committee's dismay, design engineers discovered an anomaly in their DNA construct that produced a strain of evil in the creature's soul. And, of even more concern, they discovered a propensity for many others of these creatures to irresponsibly acquiesce to this evil, that is, to go along with and assist the truly evil among them in their deeds. (Hitler, Stalin, and Mao all had huge numbers of followers, of course, to help them carry out their evil plans.)

What follows are excerpts from a long and convoluted report—*The Visitor's Report*. It is a disturbing document, one the human race will discuss and contemplate for years to come, and it leaves us with doubts about the future of humankind as we know it today.

Our attorneys sternly warned us not to publish this report, believing it will subject us to prosecution

under the harsh copyright laws protecting authors of
written material, not to mention the societal chaos
it will likely produce. However, we are convinced
this report is of such monumental importance to
mankind that it must see the light of day regardless of
consequence.

CHAPTER 9

MORE OF THE VISITOR'S REPORT

TO: All Earth Monitors

FROM: VAATU-9, Committee Vice-Chairman
Multi-galactic Research Facility

REGARDING: MODEL 25 (M25) HOMINID
DEVELOPMENT PROJECT

SUBJECT: MODEL 24 (M24) CREATURE
EVALUATION RESULTS

FILE: Project: Humanoid Creature Model 25
Design Study / Ongoing Organism Development
Project #E-3349

1. PURPOSE:

The purpose of this report is to:

1. Review the history and progress of the Planet Earth Hominid Development Project,

2. Record current Earth conditions,

3. Comment on future prospects for this experimental community,

4. List key Model 24 design flaws and how the new Model 25 design will correct these flaws,

5. Review options for the timing and means for deployment of the Model 25 to the earth environment, and,

6. Propose and review options for the mass extinction of the Model 24.

EXECUTIVE SUMMARY

Database feedback confirms that defects of sufficient magnitude exist in the Model 24 (M24) DNA master molecule to justify its replacement with an improved version, known as the Model 25 (M25). This report lists and defines both the deficiencies proposed for deletion and the advantageous features of this creature, to be incorporated into the new model. Because the M24 has over its short history far exceeded expectations, the mass extinction of the M24 will be compassionate.

Since the development and implantation of the M24, laboratory apparatus used for producing and evaluating DNA molecule design has undergone major improvements in accuracy. Your Committee is confident the proposed M25 master DNA molecule is state of the art in its performance.

The designers of the M24 are to be highly complimented on the fine design they produced. The M24 has done much to civilize itself and its surroundings, advance its quality of life, and advance the project's objectives. But much still remains to be done if overall project objectives are to be met. M24 design defects make it unlikely the creature is capable of much further significant progress.

The Committee concludes that the M24 has reached the limit of its progress and must be replaced. Five methods for mass extinction, used successfully in the past, are under consideration:

1. Planet-wide pandemics.

2. Catastrophic geographic events.

3. Cosmic ray bombardment of the planet, causing DNA mutations.

4. Extinction by the replacement life form.

5. Catastrophic climate change.

2. BACKGROUND:

The M24 creature that presently inhabits Earth was deployed in over a dozen variants to replace the obsolete M23 creature, known by Earth's current inhabitants as Cro-Magnon. At the time of its deployment to Earth, the new M24 design was a major capability breakthrough by our Species Design Section.

The author of this report, as part of this review of Earth progress, recently visited Earth disguised as an M24 of ordinary appearance, a guest of an earth-monitor in New Mexico. This earth-monitor operates an animal shelter for the observation of DNA mutations in domesticated Earth creatures. Posing as a seventy-nine year old M24, acquaintances believe my host to be a doddering old man interested in the preservation of endangered primates. As such, this observer was able to travel extensively and make inquiries without arousing suspicions. Accompanied by the New Mexico earth-monitor, visits to and questions regarding the Roswell Transit Station, and other stations as well, went unnoticed or ignored. It was reassuring to note that when M24s are questioned about the crash of our vehicle near Roswell many Earth-years ago, they scoff at the idea and tend to ignore further questions.

3. COMMENTS REGARDING M24 ATTRIBUTES AND PHYSIOGNOMIES

Flawed as it is, the M24, introduced to the planet at numerous locations some 6,000 to 11,500 earth-years ago, has flourished. There are several reasons for this. In order to improve on the earlier bipedal creature designs (Neanderthal, Cro-Magnon, Australopithecus, Homo erectus, et al.) designers made and tested many significant DNA design upgrades.

Designers found a way to substantially increase the new creature's IQ and memory, allowing it to learn better from its experiences and innovate improvements to its existence. They altered its physical structure: it is gracile, its outer skin has less hair protection, and it has a softer muscular system and more upright stance, all of which motivates it to improve its situation. Designers gave it a new larynx to improve aural communication skills, and a more sophisticated emotional makeup programmed into its control module (brain) to help it cope better with an increasingly more complex environment.

With these changes, the M24, to varying degrees, discovered within itself a determination and ability to improve its shelter, clothing, and transportation systems. Not long after its introduction to Earth, the M24 learned to cultivate grains and domesticate animals for consumption and work. These developments provided the M24 communities leisure time to improve their lives with more sophisticated goods and manufactured items, such as the wheel, and pass these on to other communities.

The M24 made the technology breakthrough we had expected of the previous twenty-three models: it discovered Earth's natural resources and how to harness fossil energy to simplify daily activities. It is believed that its technical advancements will continue on an upward path—but a path limited by fundamental DNA molecule design flaws.

A laboratory facility, conducting DNA tests of the pre-release M24 creature, failed to detect and correct anomalies (discussed in detail below) in its control module design. Almost as soon as they could communicate with each other in their earliest breeding communities, our evidence shows hominids recognized among themselves grave antisocial behaviors (evil) caused by these anomalies. To cope with them, the M24 invented stories and established organizations to set rules of conduct and behavior for each of them to follow. To add gravitas to their rules and explanations, the hominids recalled as best they could the stories of their arrival on the planet and the rules laid down by the teams of instructors who helped them begin their communities. Generations after they departed, the M24 remembers the start-up teams as gods. The M24 calls the stories of these gods and the organizations derived from them "religion." At times, these religious organizations stray from their origins, become political in nature, and cause enormous pain and suffering.

4. RECENT EARTH HISTORY

It will be recalled that, during a particularly violent conflict, one the M24 calls World War II, the outcome was in such doubt that minor interventions by earth-monitors were required to ensure an outcome favorable to our project objectives. An earth-monitor report at the time was pessimistic about Earth's future inasmuch as M24s had discovered nuclear fission[18] and were using it for destructive purposes. Earth-monitors questioned then whether the M24 had sufficient emotional and intellectual maturity to control such a force.

The destruction produced in that military conflict raised concern whether M24 civilization would decline as it had after the end of Roman influence, or, even worse, would self-destruct in a nuclear holocaust. Therefore, design began on an advanced hominid model to replace the M24 if eventually deemed necessary.

It can now be reported that conditions on Earth have improved in some respects since this last report. Notable Earth milestones:

A. The United Nations, an organization uniting the major states on the planet, was established soon after Earth's World

---

18   Nuclear fission is a process in which heavy nuclei of atoms are split apart to form lighter elements. This occurs with a huge release of energy and produces poisonous, radioactive byproducts. It is the process by which nuclear power plants produce energy today.

War II to allow nations to discuss their differences in a neutral venue rather than immediately resort to destructive force. It was partially effective during its first fifty earth-years, less so recently.

B. Significant medical advances have been made to thwart many of the major diseases that had troubled the planet. One disease, smallpox, appears to have been successfully eradicated from Earth. Polio and several others are close to eradication.

C. Evolution: Recently, within the last 125 earth-years, M24s inexplicably drifted away from religious explanations of their origin and began to use the limited capacity of all Earth's species to adapt to a changing environment, an important feature of creature DNA, as an explanation of their origin on Earth. Oddly, many M24s now foolishly believe, and strenuously argue, that they and all of Earth's creatures "evolved," after sufficient time, from nothing more than Earth's lifeless minerals into complex life forms. Given the limited sensory input of the M24 (just five senses) and its control module intellectual limitations, its actual origin is, of course, entirely unknowable and unimaginable to it.

D. A form of government that M24s call "communism" has been discredited as unworkable and, for the most part, discarded in favor of a form of government called "democracy" and an economic system they call "capitalism," based almost entirely on self-interest. Flaws in our design of M24 nature make government models based on love and sharing impractical and instead favor the use of dictatorial and/or totalitarian forms of government. We find M24s turn slothful when forced to share for the greater good.

E. M24s have continued to improve their air-breathing aircraft, and now, when long-distance travel is needed, members of the more advanced societies usually travel by air at speeds of 500 miles per hour, 30,000 to 40,000 feet above the planet surface. While some military aircraft now exceed the speed of sound, 500 to 600 mph appears to be a limit unlikely to be exceeded by civilian aircraft, at least for the foreseeable future.

F. One country on Earth, the United States of America (USA), has succeeded in visiting Earth's moon. The USA developed and used a primitive, ponderously complicated vehicle and expended huge amounts of chemical energy to overcome gravity and inertia to land two men on the moon. After an exploratory visit of several days, they returned them safely to Earth. At the end of each of their five successful expeditions to the moon, nothing remained of the contraptions that took them there but trash left on their moon and the tiny capsules encasing the men for their survival in the vacuum of outer space. These space expeditions, six in all (one failure, without loss of life), remind us of Earth's first primitive hot-air balloon travel in the late eighteenth century, which led to the development over one hundred fifty earth-years later of the jet aircraft they use today.

G. The M24 discovered, in the earth-year 1905 AD, that speed of travel for them is limited to the speed of light. Our design specifications for the experimental universe included this limiting speed and accompanying time-dilation phenomenon to confine M24 travel to its own solar system. With M24 visual sensory capacity limited to a narrow band of photon frequencies, and its mental capabilities limited to a maximum IQ of about 220 by the original DNA design codes, the M24 is not designed to cope with situations it would encounter were it to travel beyond its solar system.

H. Space travel for the M24 will not be practical unless and until it learns how to harness gravity and inertia to its needs, as it did successfully with electromagnetism in the nineteenth century and with the strong nuclear force in 1945. Several M24 scientists are making rudimentary inquiries into the effect of quantum vacuum on inertia of earthly matter, but these scientists are no further along with their studies of inertia/gravity than an M24 scientist by the name of Hans Christian Orsted was in 1820, when he first noticed a link between magnetism and electricity. It is unlikely, considering its mental limitations, the M24 has the capability to understand or manipulate the force of inertia/gravity.

I. There is concern that advancing cultural sophistication produced by the M24 will soon exceed the intellectual and judgmental capability of the M24 to cope with it.

    a. Vast sections of the M24 population exhibit below-minimum awareness of their history, governmental leaders, and government policies. This suggests they may not have the skills to compete in and cope with the increasingly complex environment they are producing.

    b. Of more concern is the M24 failure to comprehend, much less profit from, lessons of past history. Most of their current difficulties can be traced to this refusal to learn from past experience.

c. Limits we placed on the M24 intelligence level seem to be the reason for M24's difficulty in bringing to fruition safe nuclear fusion energy,[19] which was intended to provide almost unlimited radiation-free energy and to replace oil and other chemical sources of energy, which are in limited supply and pollute Earth's atmosphere. Chemical energy was made available to the M24 only as a stopgap energy supply for them to use until nuclear fusion energy could be made available. Nuclear fusion will be required on Earth soon to replace the currently used, dangerous "fission" process, which produces toxic byproducts. Our evidence suggests that because of M24's limited brain power, it may be no more possible for the M24 to develop fusion energy than it was for the previous twenty-three humanoid creature designs to exploit the planet's oil, coal, and gas energy reserves. Fusion energy may well have to await arrival of the improved creature

---

19   Nuclear fusion is the process by which light elements combine or fuse to form heavier elements, producing huge amounts of energy without the toxic byproducts unleashed by nuclear fission.

design, the M25, scheduled for implantation in the near future.

d. Although minor military actions continue to occur on Earth, the fear of nuclear detonations has prevented major outbreaks of armed conflict until recently. Unfortunately, the latest efforts by major powers to prevent proliferation of nuclear weapon capability are unraveling. Suicidal religious lunatics have gained control of significant economic entities, have acquired vast arsenals of nuclear, biological, and chemical weapons and have publicly declared they intend to use them against their perceived enemies, believing extermination of life on the planet preferable to cohabitation with what they call "infidels." The Committee believes that without intervention the planet soon will be incapable of sustaining life, thus ending the experiment.

5. M24 DESIGN DEFICIENCIES / REMEDIES IN THE MODEL M25

The M24, with an advanced control center giving it the ability to modify its environment to improve its life, has for the most part civilized itself and its world. However, the M24 itself has not evolved as fast as have the surroundings it has transformed. Many of the M24 design characteristics, needed early on for the model to emerge from hunter-gatherer status and civilize itself are no longer applicable to its present world. Moreover, evidence confirms that certain of its original characteristics, thought of early on as mere design flaws, now stand out as gross design deficiencies.

The deficiencies listed below have been corrected in the new M25 (rectifications are shown italicized) and are expected to mitigate or eliminate the evil and concurrent acquiescence we find in an unacceptable percentage of M24 souls that have returned to the master database.

A. Original M24 design specifications called for anomalies to occur in no more than 0.05 percent of the M24 population. Unlike other, less complex Earth creature populations, which meet their design specifications and deficiency limits in this regard, M24 designers, in their efforts to bring their project to fruition on schedule, missed meeting this particular specification requirement. Evidence shows approximately 20 percent of the M24 population is afflicted with various physical, mental, and conduct aberrations, some requiring incarceration to protect the larger population from harm. Some of the more important aberrations are:

   1. Deficient emotional control, leading to gratuitous acts of violence, such

as murder, robbery, assault, and rape;

2. Susceptibility of individuals in positions of authority to corruption, hubris, narcissism, and lust for power, leading to vast misuse of their authority;

3. Mental illnesses and deficiencies causing abnormal behavior and emotions.

B. The M24 was implanted on planet Earth in multiple variants, each with equal performance capabilities but with distinct variations in facial and skin features to allow evaluators over the centuries to identify where on the planet they were originally introduced. Without the ability to identify the original location of each M24 colony, evaluation data is skewed and invalid.[20] DNA programming encourages the M24s to form tightly knit societies (tribes) for protection while emerging from hunter-gatherer status, to reject other M24 variants, and, even more disastrously, to exterminate models 19 through 23. Much unanticipated turmoil has resulted from this programming feature.

The M25 will arrive in singular form—no variants. Its facial and skin appearance,

---

20 Author's note: For an excellent, well-researched review of how geographical location affected the rise of civilizations, read *Guns, Germs, and Steel: the Fates of Human Societies*, by Jared Diamond.

a composite of the M24 variants, is configured to convey a trusting and caring nature to the M24 variants with which it may interface.

C. Unlike skin, bone, and muscle, the M24 is unable to repair damaged brain and nerve cells, leaving nerve-damaged creatures dependent on others. Until recently M24s treated injuries with unskillful medical attention, mercy killing, or death from neglect. Now improvements in medical care, along with M24s aversion to mercy killing, have resulted in huge numbers of semi-functional M24 creatures living extended, unproductive lives. The M24 is attempting, with only limited success, to develop its own means to repair damaged nerve and brain tissue.

The M25 DNA design provides more comprehensive tissue regeneration and self-repair. Nerve, tendon, eye, ear, and brain tissue will regenerate if damaged, as does M24 skin, bone, and muscle tissue (and as do tails on some lizards and eyes on conches).

D. Genetic intelligence, averaging (by definition) IQ100 and considered largely responsible for the emergence of the M24 from prehistory, is now an impediment to further M24 developmental progress in the complex culture it has developed for itself. Significant numbers of the M24, approximately 10 percent of the population, are unequipped or unable to cope with modern life today and require assistance just to survive.

IQ and intuition levels, and capacity for memory and recall, are increased 40 percent in the M25, enabling it to cope with the complexities of the M24 environment. The M25 IQ bell curve has been sharply right-skewed to limit the number of individuals with marginal ability to cope. Below an intelligence quotient of 95, few M25 individuals could manage, in any meaningful way, life's challenges in the complex world that it will evolve, or, for that matter, in the civilized environment that has already evolved. M25 fetuses with IQ levels below 90 will be stillborn.

E. The M24 aggression level was established when the creature needed to survive in an untamed "kill or be killed" habitat. This point is too high for a complex bipedal creature living in an advanced, civilized culture and environment.
   The M25 aggression set point is significantly lowered to reduce inappropriate aggressive behavior, and combined with this reduced aggression level is a commensurate increase in the empathy quotient toward its fellow creatures.

F. Masses of the M24 creatures tend to follow charismatic leaders and outdated ways and customs, and follow many practices without adequate concern for their own individual best interests. Herd instinct, once necessary to help the M24 survive in a primitive world, is too strong for its own good in a modern civilized world.

The M25 instinct to form large social groups and mindlessly follow charismatic group leaders is reduced compared to the M24.

G. M24 verbal and visual communication skills are far too primitive for the modern world it has produced. Spoken communication is very inefficient for conveying and accurately describing thoughts and is susceptible to manipulative propaganda. Written communication is often unclear and subject to misinterpretation. Records requiring maximum clarity, such as legal documents, are carefully and expensively produced and are understood and interpreted only by those skilled in the legal profession. Untruth is too easily disguised as truth.

    1. The M25 verbal communication skills are enhanced. The Broca's area of its brain is enlarged, and the position of the larynx slightly repositioned and enlarged to allow for a broader range of vocal sounds. The M25 is programmed to favor English as its primary language because it's easy to learn, is widely used, and has considerable tolerance of error. The M25 will also use whatever is currently the dominant local language (spoken by M24s in the geographical areas to which an M25 is assigned) as its secondary language. Within five to six generations, it will evolve a more efficient language of its own.

2. Devices such as digital-image editors and Photoshop-type computer programs, which edit and alter visual images to suit the author's purpose, now bring the authenticity of visual communication (photographs, cinema, TV, video, etc.) into question. The enhanced IQ of the M25 improves its ability to see through this form of subterfuge.

H. The level of pain experienced by the M24 was appropriate for its transition from primitive to civilized existence, but it is now judged too high for a civilized world. The M24 developed advanced medical procedures and established excellent facilities and medicines for repairing itself, thus severe agony is no longer required for diagnosis or to limit bodily activities while the M24 creature repairs itself.

The M25 will not experience acute pain. Pain for the M25 is limited to discomfort levels that cause it to reduce its activities and seek medical attention.

I. Body-weight control mechanisms for the M24 were intended to allow for storage of energy for use in times of deprivation. Availability of food has increased so substantially that concern for food deprivation is virtually nonexistent in the developed areas of Earth, and too many hominids today are overweight.

The DNA of the M25 is modified to limit its body weight to no more than 15 percent over specification.

J. The M24 is concocting chemical and mineral formulations unimaginable when it was designed and released 11,500 earth-years ago, and these formulations are now overwhelming its body's natural defenses. Cancers and birth defects form in the M24 at an unacceptable rate. Considering the limited intelligence of the M24, mechanisms developed by it to fight cancers and birth defects are likely to be marginal at best.

Incorporated into the M25 DNA are stronger natural body defenses against cancers and intrusions into the body of unanticipated chemicals and compounds, including an improved immune system defense against microbial and parasitic diseases.

K. Data shows M24 births involve an unacceptable rate of miscarriages, premature births, low birth weight, congenital malformations, maternal deaths, and other defects.

The design specification for the M25 limits all birth defects to less than 0.025 percent of births, no exceptions, a rate half the acceptable rate for all other hominid defects.

6. INCIDENTAL COMMENTS

A. It is noted that M24 populations are reacting with concern to the occurrence of what they term "crop circles," which the Committee knows are produced by some of our less adult representatives in their leisure time. The M24 has invented interesting but incorrect theories as to how these crop circles are produced, and although they don't seem overly upset by this minor damage to their crops, it is certainly annoying to them and costly to the affected landowners. These crop circles do nothing to advance project objectives and suggest to your Committee that the home office may well have assigned too many project engineers to this experimental site (Earth). We suggest either that these representatives spend less time tormenting the M24 population and devote more time responding to individual M24 mental requests (prayers), or that they should be reassigned to other projects.

B. Soon after the last report, some of our Earth visitors became careless in their selection of sensory frequencies that hide their entry into Earth's atmosphere from the view of the M24. Beginning in the Earth year of 1947, thousands of these vehicles have been observed by M24s. They were at first extremely agitated and termed these vehicles "Unidentified Flying Objects" (UFOs). Due to their rounded exteriors, some on Earth call them flying saucers. A few of our less serious visitors found this humorous and have, while observing M24 activities and approaching our entry sites, unnecessarily continued to irritate M24s by intentionally displaying their vehicles in photon frequencies visible to the M24. Fortunately, except for one error at the Roswell Earth Transit Terminal, which resulted in the loss of four of our engineers, the vast majority of monitor vehicle pilots have been careful to select sensory frequencies invisible to M24s before beginning a final approach and landing at Roswell and the other terminals. All Earth visitors should be mindful of M24 sensibilities and should severely limit visibility over the planet while observing Earth activities.

7. GENERAL COMMENTS REGARDING SPECIES REPLACEMENT

The author of this report looks forward to a timely completion of the new creature design (Project E-3349) now that the recommendations contained herein have been incorporated into its new DNA molecule. The Species Design Center believes it has addressed all the above design deficiencies and will soon be releasing a much-improved creature model to Planet Earth.

The designers of the M24 are to be highly complimented on the fine design they produced. The current design has done much to civilize itself and its surroundings, advance its quality of life, and advance the objectives of this project. But much still remains to be done if overall project objectives are to be met. It is unlikely the current M24 design is capable of much further significant progress.

In the opinion of the Committee, the M24 has reached the limit of its progress and must be replaced.

8. OPTIONS FOR MASS EXTINCTION OF THE M24

Your leadership is in the final stage of determining the best means to eradicate the M24 from planet Earth and replace it with the M25. In the past, when target creature populations reached the end of their useful life and were judged unable to produce further meaningful contributions to their cultures and infrastructure, the methods listed below were used.

A. Introduction of planet-wide pandemic disease(s) fatal to most if not all the target life forms. Target populations are reduced to an unsustainable size and soon disappear.

B. Catastrophic geographic events, such as massive and widely distributed volcanic eruptions, violent meteor impacts, earthquakes, floods, etc.

C. Introduction of X-rays or other appropriate cosmic rays causing DNA mutations in target creature reproductive systems, resulting in sterility and vastly reduced inclination for sexual reproduction.

D. Extinction by the creature's replacement life form.

E. Climate change that exceeds the target
   creature's ability to adapt.

These methods have worked rather well
before, as in the case of the dinosaur during
what is known to Earth's paleontologists as
the Cretaceous-Paleogene event. But now, with
M24 culture and infrastructure extending
around the planet, its extinction will require
careful planning, so as to minimize disruption
of its valuable infrastructure and to allow
the M25 to assume operational control.

Your leadership has taken into account
the following considerations regarding each
of the above listed techniques used for mass
extinctions.

Method A (worldwide pandemic disease[s])
is still considered a viable alternative for
reducing the M24 population sufficiently to
prevent it from rebreeding into a sustainable
life form. Some form of a fast-acting leukemia
would probably work best. When the M25 assumes
primacy over the planet, it will find the
residual M24 population as unfit for much other
than slavery or menial labor. It is understood
that as soon as the M24 detects the onset
of a pandemic, it will immediately instigate
intensive medical research into means for
defeating or severely limiting the effects of
the disease. Rough calculations indicate there
is no more than a 21.6 percent chance that the
M24 is clever enough to defeat such an attack.

Other than meteor impacts, **Method B** (catastrophic geographic events) is not being seriously considered at this time, because of the damage that would be inflicted on existing culture and infrastructure. When used in the past, this technique was very effective. Meteor impacts could be utilized if other means prove unworkable, but damage to the structures and facilities produced by the M24 would have to be limited to the immediate area of the impact—let's say within a five-hundred-mile radius. The unintended consequences of this technique might well exterminate too many of the untargeted, less complex species needed to sustain the M25. This method is considered at this time as a last resort.

Method C (catastrophic introduction of X-rays or other appropriate cosmic rays) is under serious consideration, because it would be painless to the M24, and it is unlikely the M24 medical profession could counteract the effects before extinction is complete. Hominid replacement ratios in the more culturally advanced societies of today are below replacement requirements, but in the less advanced cultures, births far exceed deaths. Study is underway on how to prevent or limit DNA mutation of untargeted creatures, and to minimize other collateral damage.

Method D (extinction by the creature's replacement life form) is not under consideration, because the M25 is not programmed to attack hominid creatures they perceive to be foreign to or "different from" their own species. This technique worked when follow-on creatures were introduced up to and including the extinction of Models 19 through 23 (the Neanderthal, early man, and Cro-Magnon models).

Method E (catastrophic climate change) is not under consideration, because the M24, with its complex culture and scientific capability, could easily defeat the attempt. As well, climate change sufficiently catastrophic to eradicate the M24 would also eradicate most of the other life forms the M25 will depend on for its survival.

NOTICE!

Every consideration has been given to the peaceful introduction of the M25 Advanced Hominid Design onto Earth and the transition of those remaining M24 variants to menial tasks. Nevertheless, you are hereby warned that the current M24 is programmed, as were previous hominid designs, to attack and exterminate intelligent hominid life forms foreign to or different even in subtle ways from its own makeup; it may, when it comes to realize its fate, attempt to exterminate M25s, wherever it finds them. Once aroused, it will use all the powerful weapons it has accumulated. Project supervisors are warned to have countermeasures for each of the significant M24 weapon systems when the M25 is ready for introduction to Earth.

Failing adequate protections when the M25 is introduced to Earth, project supervisors are likely to find it necessary to reeducate, relocate, enslave, or militarily exterminate the current living M24 population (much as one M24 variant did to another in the sixteenth century, upon arriving on the newly discovered shores of the South American continent).

EARTH-MONITORS ARE URGED TO RETURN THEIR COMMENTS REGARDING THIS REPORT TO THE COMMITTEE NO LATER THAN ONE WEEK FROM RECEIPT, SO THEY CAN BE ADEQUATELY CONSIDERED BEFORE THE UPCOMING MEETING.

End Of Report

What you've just read is an abridgment of a long and detailed report written by an alien visitor to our planet. Some readers will reject it as the musings of a demented author desperate for attention. Others wedded to scientific endeavors that proscribe references to supreme beings for explanation of phenomena will reject it outright.

Still others, however, will note that it seems to answer a great many, perhaps all, of history's most profound questions, questions that had been fully answered by theology until the advent of the *scientific method*, a study technique that evolved five hundred years ago and which requires—mandates without question—that resort to a supreme being for explanation of phenomena is expressly forbidden under threat of ridicule and expulsion from the scientific community. For the educated, open-minded reader willing to accept its content, this report provides answers to questions such as: Who are we? Who and what is God? Why are we here? What is to become of us? It also appears to explain phenomena such as crop circles, UFOs, and other sightings that have intrigued mankind since the dawn of history. And finally it puts to rest once and for all the dispute between evolutionists and proponents of intelligent design. Intelligent design wins.

Before the reader rejects this report, he should carefully examine how it, and the arguments supporting it, explain, with rigid logic, his own world and how life arrived on Earth. The reader should ask himself if, as all authorities that have seen it believe, it might well be legitimate. Ask yourself if you're still secure in your belief that humans are masters of their surroundings.

After reading this report, will any of you wonder whether your Maker is displeased with you, and might well be considering hominid design improvements that supersede you—that, in effect, you are going to be ... "*THROWN AWAY*"?

~Elliot Flindecker III, staff writer

PART II

CHAPTER 10

Past and Future

By now some of you surely have begun to suspect that the "scenarios" I described in past chapters may be more than just scenarios. How could something as specific as *The Visitor's Report* be a scenario? Good question. Up to now I have been using your sense of logic to lead you forward—hopefully, with success. But to continue I will need to introduce you to a way of thinking about what you see around you: the concept of "past, present, and future."

Most of you think of the "present" as a period of time, perhaps an interval of time that surrounds you with all that you see, touch, smell, and hear. Memory of very recent events is still for you the present. Your home, its contents, your car, and all your friends are your "present."

But in actual fact there is no "present," only past and future. The present is nothing more than an infinitely thin membrane separating past from future. Take a photo of a friend or scene and you are instantly looking at the past. Your photo is showing you a past etched in granite, immovable, unchangeable. What you photographed is a past as unchangeable as the sphinx or the pyramids of Egypt. Words you uttered seconds ago are in your past. No matter how unintentionally you may have spoken hurtful words and how much you wish to retract them, they are as much a part of the past as the Washington Monument in Washington, DC.

A UFO time traveler, venturing into our future by means of time dilation and "photographing" or otherwise recording the future of any sector of the universe, appreciates that he is recording a continuously changing, malleable scene slowly evolving in infinite ways by means of the trillions of micro-decisions made independently by seven billion human brains, along with Earth's uncountable creatures and plants, in the split-second "present" of their lives. Our UFO time traveler is wise enough to understand that the future he sees and records is *likely*, but not certain, to occur if human activities continue on the path they are on. He also understands that as he travels further into the future, the picture he sees from the window of his transporter becomes less specific, more blurred, until at last, like long-range weather forecasts, the picture becomes virtually useless.

Let me invite you to board a UFO with me for an exciting voyage through this thin membrane separating past from future and see what is in store for the human race. I must warn you, though, that if you join me you will need a strong grip on your senses, for I made this voyage once before and must confess I was not prepared for what I saw.

Knowing human frailty and stubbornness for what it is, my hosts for our travel to the future calculate there is less than a 3.2 percent chance the future you will see changes in any significant degree, that the M24 will suddenly come to its senses, end its march toward nuclear war, and make serious efforts to develop inexpensive nuclear fusion energy. And, of course, only if these drastic changes are made in the very near future will the Committee put off its plans for human extinction.

* * * *

In early May, 2030, with the unexpected media release of VAATU-9's report, the M24 replacement program became urgent for the Committee. They acted quickly and called a meeting of earth-monitors to present their program and get earth-monitor input.

On June 30, 2030, while I was still visiting my friend Harry Bates, he and I got word of the date and time of a special meeting we were to attend. To get to it we were to meet people at the Roswell airport and proceed to the meeting with them. Travel arrangements were all arranged for us.

Harry lives on the north side of Nogal, New Mexico, with his darling wife, Sue, about fifty miles from Roswell. The morning of the meeting, Harry's car headed east on route 380, not another car in sight. Our ninety-minute trip to the Roswell airport was uneventful.

(I was not surprised later to learn that an unmanned aerial vehicle that had circled at thirty-five thousand feet over Nogal for the past twenty-seven hours gently moved out of its orbit as we departed Harry's home and assumed a course of 095 degrees, paralleling route 380E. The pilot of this vehicle, seated comfortably in his Albuquerque base cubicle alongside his young female copilot, guessed the route Harry and I would take and allowed his vehicle to drift ahead of his target. Our car, as was usual for Harry, exceeded the speed limit by about ten to twelve miles per hour, just enough not to be noticed by speed radar. But it was never out of view of our Albuquerque observers. By the time Harry's car entered the airport parking lot, our watcher was orbiting overhead and downloading, to a large plasma screen on the wall, a high-definition picture of Harry and me emerging from his car, locking it, looking around for official notice, and then heading toward the terminal, Harry's briefcase in his hand. Harry and I disappeared from the overhead vehicle's view into the terminal. Its task now over, the picture switched off, but the vehicle maintained its silent orbit overhead, pending our return to the car. Its controllers returned to their discussion of where, when they got off duty, their assignation for that evening would occur.)

Inside the terminal, travelers were scattered about, some absorbed in news magazines or their own personal computers. Almost none even looked up as we entered. One woman, glanced our way as we passed her, but immediately went back to her hopeless task of corralling a wayward child.

Harry and I entered the men's room and disappeared from the view of the terminal passengers. Four men met us there, two of whom hurriedly exchanged jackets and hats with us, quickly made minor changes to our facial appearances, and took Harry's briefcase. Then, accompanied by the other two, we departed the men's room and walked to an exit from the terminal to the tarmac. Once outside, we followed our friends to an elderly, nondescript Piper Aztec twin-engine, propeller-driven, six-seater aircraft standing near several other private planes lined up across from the tower. To any observer this would look like an ordinary business trip by low-level executives. Nothing about it was worth a second look.

Harry's and my lookalikes left the men's room three minutes after Harry and I departed and walked to the ticket desk and asked some perfunctory questions, then, with Harry's empty briefcase, the two went to Harry's car, started its engine, and headed out of the parking lot and back to Nogal.

(An annoying alarm interrupted the intimate discussion between the Albuquerque UAV pilot and his copilot before they could decide in which of their condos they would spend the night. Sensors on their seventy-two-million-dollar vehicle circling the Roswell airport were indicating that Harry's car was leaving. The two pilots reluctantly returned to their day job and followed it home.)

Harry was directed to a seat in the cramped third row of the Piper Aztec. One of our associates sat next to him, and the other sat next to me in the second row. The pilot then entered the plane, settled himself in his seat, and strapped himself in.

Finally a sixth person, whom I could see only from his back, entered to sit next to the pilot up front. When all occupants of the plane were secure in their seats, the engines started and the control tower directed the pilot with instructions for taxi and takeoff. It would be a flight conducted under visual flight rules with no notice given of the plane's intended altitude or destination.

Once we were airborne high in the overcast New Mexico sky, air traffic control instructed the pilot to tune his transponder to a given frequency so they could monitor the flight. For about fifteen minutes into the flight, the transponder answered the usual repeated ground-based requests, "squawking" its location and altitude, and then the pilot abruptly stopped its replies. Without signals from the transponder to ground-based facilities, I knew the plane's altitude would be impossible for ground control to know, and its location only faintly noted, due to the mountains in its flight path and its low altitude. Soon we began to lose altitude, and now even the plane's location would disappear from the air traffic controller's radarscopes, providing them some concern but not enough to call for any kind of a search. I noticed that their communications to our plane were now going unanswered; we had changed to a new radio frequency.

When our pilot determined his location had been lost to ground-based controllers, he made an abrupt right turn to a course of 010 degrees. Before long, the plane descended into a valley surrounded by high mountains, until at last I could see the wing flaps lower and hear the landing gear extend. A strip of scrub grass appeared ahead, and the plane settled onto it as the engines quieted to idle. Settled on the ground now, the plane taxied to a gas pump near a small hangar and stopped. The pilot shut down both engines and motioned us to depart.

When the four of us in the back of the plane got out, our associates gathered about us and provided us our first inkling of what was to take place.

"You will be attending a conference called by VAATU-9 for his earth-monitors. He expects, of course, that each of you has read through his recent report and will be contributing your opinions and comments about it to the Committee members who will be there in virtual attendance." That was it. There were no other comments or explanations.

We followed the men to a rusty door constructed into one side of the large hangar door. The man in the lead unlocked it, pulled its handle with both hands. Like many such rusty hangar doors, it opened it with a screech. We entered a gloomy hangar containing just three planes: a Beechcraft Bonanza, a Piper Cub, and, of particular interest, a classic Beechcraft Stagger Wing dating back to the 1930s, the golden age of flight. It was a classy airplane with a brand new cloth covering—obviously loved by its owner. We proceeded to a door in the back that opened to a small room. Its lonely overhead light illuminated a dusty office equipped with two file cabinets, one with its top drawer partly open; an old, scarred desk; a chair; and a telephone, obviously seldom used. An empty five-gallon paint can, top removed, served as a wastebasket. Yellowed pictures of Presidents George Washington, Lincoln, and Ronald Reagan adorned one wall. The other walls were bare, even of windows.

To the left side of the room were thirteen steps leading down to a door, which, when unlocked, opened to reveal a dimly lit tunnel leading, it seemed to me, toward the mountain to the rear of the hangar. We entered the narrow tunnel in single file, walking on loose concrete slabs purposely warped to noisily announce our presence in the tunnel. After what seemed to be at least several hundred yards, our group came to a sliding door leading to a large, well-lighted room.

When I was able to reach the man in the lead, I whispered, "I have a question. Who was the gentleman sitting beside the pilot? Somehow he looks familiar."

"We saw no need to introduce you. His name is VAATU-9."

CHAPTER 11

THE MEETING

VAATU-9, Harry Bates, and I entered a brightly lit conference
room and looked around. Nearly all the earth-monitors
had already arrived and were in private conversations among
themselves. We immediately joined the discussions with fond
acquaintances from around the world.

All earth-monitors speak nearly perfect English and several
other languages in addition to the tongue native to the country
to which they are assigned. Most in attendance wore clothing
common to their assigned locations. Mixed in with formal
business suits in American, European, and Asian styles were a
few African national costumes and Middle Eastern robes, so
the room took on the flavor of a United Nations meeting, but
with none of the antagonisms and hatreds endemic to that
organization. Here we were all friends. Our assignments on
Earth were to blend into society and report on the happenings.
Nothing else. In no way were we to alter the natural flow of the
world's social activities.

Earth-monitors look forward to these irregularly scheduled meetings, particularly the banter before and after the official portion of the business meeting. All of us are selected for our photographic memories and gregarious, outgoing personalities, precisely to encourage acquaintances to share their experiences and opinions. Many earth-monitors have been on their assignments on planet Earth since before the American Civil War, so they bring to any discussion an unusually broad perspective toward life and world activities. A few, new to their assignments, busy themselves getting advice from the elder monitors. Conversations at these meetings are fascinating and difficult to end no matter whom one turns to.

When the last of the 130 earth-monitors from around the planet arrived, we were asked to find seats. Conversations died down as we looked around for somewhere to sit. Finally, when we were seated, seventeen Committee members made up of system managers, DNA-design specialists, and assorted engineers from the experimental laboratory overseeing Organism Development Project #E-3349, filed in from the right and took their places at a long table set before a blue curtain, on which was affixed a large white screen. All 130 earth-monitors stood when the Committee entered and remained standing until each Committee member was introduced and seated. VAATU-9 sat, prominently, near the middle of the group, and, significantly, to the Committee Chairman's immediate right.

Other than VAATU-9, who had undergone the rigorous process that converts eleven-dimensionals into three-dimensional beings, and who had retained his name in a format common to eleven-dimensionals, none of the Committeemen we saw at the table were actually present. They were there in a virtual sense, utterly real to the eye, thanks to sophisticated 2030 holographic technology. Video cameras and microphones, arranged around the room, transported everyone holographically back to the eleven-dimensional world of the Committee members with similar realism.

The Chairman, a slim, distinguished man with expressive, penetrating eyes, opened the meeting. "You have all read the very thorough report prepared in fine fashion by VAATU-9. I congratulate him on a job well done," he said, motioning to his colleague on the right.

When the applause for VAATU-9 died down, the Chairman continued. "By now you have, I'm sure, digested what this report says and what it implies for the future of our grand experiment. It tells you what has been decided and what little work remains. You are here to give us your guidance as to how best to discard the M24 in favor of our new M25. We value your opinions and input because you are so intimately involved with the daily life of the M24. We are sure any personal attachments you may have made with these M24 individuals will not cloud your thoughts or judgment. We want your best, unbiased input. I know you will provide it over the next few hours of our meeting.

"To begin, we will review the background and reasons for bringing forth an improved model hominid. As you are all aware, the M24, despite its extraordinary achievements in bringing itself out of prehistory, exhibits side effects that are painful for all of us to watch. We see from your reports that the overall defect rate among the M24 approaches 20 percent, far exceeding the 0.05 percent rate we established as a target, and which we have achieved by and large in nearly all the less complex creature categories. Deficiencies, in addition to the dreadful tendency toward iniquity, range from birth defects, emotional problems, cancers, mental diseases, to sexual deviancy. About 2 percent of the population engages in criminal activity against other individuals, especially family members, and, more disturbingly, outright criminality— murder, robbery, torture, mutilation, mental cruelty, and the cultivation and distribution of dangerous drugs. The list of M24 depravities is endless. Our database shows—and without it we would not believe—that some deviant M24 creatures actually find the bloodcurdling shriek of victims in mortal agony thrilling. They call it *sadism*—an anomaly that is impossible for us to imagine or allow to continue. Let's face it, everybody. In the vernacular of a US Marine I once met, 'We fucked up!'"

The Committee Chairman waited to allow his comment to sink in. When the buzz died down, he went on. "The tendency for hostility toward others who appear different from themselves, a characteristic we instilled into them so that they would form groups, for self-protection, has degenerated into a propensity for ghastly, vicious, large-scale wars using every weapon their fertile minds can invent. Far too much work has gone into this experiment to allow the M24 to render our planet unfit for any form of life whatsoever.

"Most of you earth-monitors are unaware of the arduous procedures involved in identifying all the physical constants and characteristics required to sustain three-dimensional life. Our early trials showed us that the location selected for our experiments must exhibit forty-one very special, precisely calibrated characteristics. Four energetic forces—electromagnetism, weak nuclear forces, strong nuclear forces, and gravity, all sized and precisely balanced—allow three-dimensional life to exist in our new universe. Stringed, one-dimensional particles of energy, combining in a multitude of ways, make up all the necessary subatomic particles in the universe that provide solid mass for our new universe.

"We found the total mass of this universe had to be held to tight tolerances to allow elements heavier than hydrogen and helium to coalesce during the initial explosion. Supernovae must detonate early in the history of our new universe at precisely the right time to produce the heavy ash that brings the heaviest elements to our special planet, but must not explode later on and destroy living creatures after they are introduced to our planet. White dwarf binaries produce the one element we had trouble with: fluorine.

"Our experiment's special planet had to be located near a precisely sized star, one small enough to produce a uniform flame yet large enough to allow us to position our planet far enough back from it so tidal forces would not disrupt the twenty-four hour rotation period needed to hold planet temperatures within life-sustaining limits. Furthermore, the planet we wanted needed to be large enough to retain water yet small enough to throw off toxic amounts of ammonia, methane, hydrogen, and helium from its atmosphere.

"And our tests even showed the sphere we selected had to have a sizable moon to stabilize its rotation axis at an angle of twenty-three and a half degrees to produce seasons for plant growth. An oversized planet in an outer orbit, to deflect incoming asteroids and other celestial bodies, would be a big plus. All of these characteristics had to be held within plus or minus a million-billionth of a percent in order to achieve our aims—a tall order indeed, requiring our best minds.

"But enough of our past history. Let's get on with the meeting and discuss in detail why we are here. To do that, I will introduce our new Director of Quality Assurance to you, who will discuss the quality problems plaguing the M24 ever since its first implantation."

The new director, a short man with rounded shoulders and a scowl etched into a pale face, stood and, leaning forward as if deep in thought, walked to the lectern. When the polite applause for him died down, he adjusted his black, heavy-rimmed glasses and began his presentation.

"Our number-one goal for the M25 was to correct the most egregious M24 defects, not to redefine the creature completely. Fortunately for us, the M24, according to your reports, does exhibit many useful traits, so we have been able to base our new design on the M24 DNA codes. We did not have to start from scratch.

"We developed sophisticated computer simulation procedures to evaluate the M25. Our first effort delivered many of the improvements we wanted but also more deficiencies than we were willing to accept. So we backtracked, made major refinements, and designated our revised laboratory prototype the XM25-1. It was not long before our simulation techniques had improved even further, allowing us to detect anomalies we never suspected could infect our new species and to discover several more hitherto unknown. As a result, we returned to the design phase for further revision, and the XM25-2 finally emerged. Tests of this prototype creature's DNA code were extensive and, I must say, exhaustive.

"The latest tests demonstrate that all defects, including criminality, birth defects, sexual deviancy, and mental illnesses, to name just the more serious, will total less than 0.05 percent of the population, bringing it into line with the defect rate of the other animal species. Major military conflicts, minor military conflicts, and violent crime will be things of the past. As a matter of fact, we expect history books written after the advent of the M25 will make for dull reading, containing mostly important achievements produced by the M25 in science, medicine, and political science and almost nothing about military conflict.

"We further predict a significant reduction in what we call 'social friction,' due not only to the enhancements already mentioned but also to revisions we made to the M25's nervous system, brain, and extremities. Nearly the whole organism is fitted for self-repair, and this, along with the strengthening of the creature's immune system, will drastically reduce its need for medical assistance and convalescence. It is interesting to note that increasing its intellectual capacity, along with all these other improvements, required only a 7.23 percent increase in its cranial case volume.

"We think this is a major design breakthrough that will enable our new creature to make major cultural enhancements to its life on earth. In sum, the M25 will finally fulfill our project objectives. Our new test techniques and computer equipment confirm that our latest species will be a vastly improved, much less defect-prone creature. We certify it is ready for release. As earth-monitors, however, you will have to be especially vigilant when the M25 appears and must immediately report any sign of defects."

It dawned on several of the earth-monitors as we listened to this presentation that the very same Committee had been equally certain of complete success when it introduced the M24 long ago.

CHAPTER 12

METHODS

The Chairman rose, thanked the QA director for his comments, and said, "I want you all now to turn to Section 8 of VAATU-9's M24 evaluation report, where mass extinction techniques Methods A through E are discussed. Each of you should have a copy of the report in front of you."

A hand rose in the center of the group.

"Yes?"

"Before we begin discussion of Methods A to E, may I ask a general question regarding the project?"

"Proceed, but keep it brief. All issues pertaining to the body of the report and decisions resulting from it are set. Only the method of extinction of the M24 is yet to be decided."

"Just one question, sir. Some of us earth-monitors have married an M24 and have families from them. How will those M24 families be affected by the decision to terminate?"

"At the time of your deployments to planet Earth, you all were warned not to form deep emotional attachments with the M24, and we assume that none of you are anything more than superficially attached. Your M24 family members will be subject to the extinction method finally selected. The one thing we will allow is that judgment of their souls upon delivery to our database will be favorably considered. You are all aware the world you inhabit is nothing more than an experiment—one with which nothing can be allowed to interfere. You knew this before arriving at your posts.

"Any other questions of a general nature?"

A hand rose in the front row. "Several of your methods call for disposing of the M24 while the new humanoid species is being introduced, a technique you call 'concurrent implantation.' How do you see concurrent implantation happening? According to your species specification, the M25 facial appearance is a composite of various M24 variants, and it will be recognized by the more observant M24s. They will see the difference. Won't the sudden appearance of M25s in M24 communities cause alarm and arouse authorities to question their presence?"

"Good question," said the Chairman, "one to which we've devoted considerable effort. This has been a big challenge for us because, as you know, our previous implantations were rather uncomplicated. They involved placement of just fifty- to two-hundred-member clans into isolated primitive breeding communities. We chose sites along rivers, surrounded by lush vegetation, and in warm, dry climates, where only minimal skills were necessary to survive from day to day.

"Now, with concurrent implantation, we insert the M25 into functioning communities where the obsolete model is declining but nevertheless still quite active. To do this we must and will surreptitiously introduce records for each M25 into the vast record files already existing around the planet so the M25 can meld into its society quietly and without fuss. We use techniques similar to those used by the US Marshals office when relocating protected witnesses in their Witness Relocation Program. We produce records of past histories down to the tiniest details involved in personnel placement. Finally, and very importantly, the new adult M25 is preprogrammed with a vague, false memory of an early life, along with an education in language, customs, skills in the various occupations for its locale, and social skills appropriate for the community into which it is introduced.

"For the local M24, arrival of M25 families will seem a perfectly normal occurrence. They will welcome them as they would any new arrival. They may wonder about the 'different look' and superior intellect these new arrivals apparently possess, but they should not be concerned with much else, as far as we can see. We expect the M25 soon to become valued members of the unsuspecting M24 communities around the world.

"Within two generations, the M24, no doubt, will begin to suspect something is wrong, and alarm will spread. It will soon be noted, for example, that the inevitable sexual contact between the M24 and the M25 produces only sterile offspring, just as horses and donkeys produce only sterile mules.

"But the factor we expect will most alarm the M24 is the impression that the M25 usually seem to find the better jobs and better opportunities, seem to live in the better communities, seem to achieve better scores in school and university, and generally appear superior to the M24. It won't seem fair to them. In two generations, maybe less, the M24 will notice it is being relegated to the more menial and dead-end jobs. We expect, but cannot guarantee, that by the time the M24 population recognizes its fate, it will have diminished sufficiently in numbers and influence to be unable to pose a lethal threat to the M25. There will be localized violence and assaults, to be sure, similar to those that at one time targeted Jewish communities perceived to be different and 'superior,' but this violence should diminish with time. This minor violence is the price we pay if we decide to select concurrent implantation of the M25."

A Committee engineer added, "Part of the consideration of the concurrent implantation method is to figure out how to produce a truly massive, simultaneous introduction of the M25, perhaps in the hundreds of millions. We will need to greatly expand the number of transit ports and earth-monitors to man them. All of the transit ports, of course, will remain invisible to the M24. Each facility will provide an entrance point for the newly minted M25 and its family, including children, aunts, and uncles."

VAATU-9 stood. "Other methods of M24 termination listed in our report utilize 'consecutive implantation,' i.e., the M24 is completely erased from the planet before the M25 is introduced. This technique addresses, of course, the concern some of you have expressed regarding the transmission of toxic thoughts from one species to the next. We used this technique in earlier times, but it, of course, carries drawbacks if we were to use it today.

"Any other questions before beginning discussion of Method A?" There was a low buzz of discussion, then silence.

On the screen behind the Committee appeared: **Method A: Planetwide Pandemic Disease.**

A hand went up in the back of the room. "Sir, I think a disease could be formulated that could not be controlled by the M24 before all of them passed on. It would be a disease based on rapidly mutating microbes similar to influenza and AIDS but modified to mutate more quickly and pass from one to the other of the M24 by inhalation. The microbe would be far more lethal and quick acting than the influenza pandemic that occurred after the M24's World War I, and, hopefully, less painful during its active period.

"I have contacts back in our laboratory who have experimented with just such microbes and they report the job is doable. The microbes they envision will spread across land masses in a matter of six to eight weeks, and in no more than twelve weeks the infrastructure of the world would be devoid of living M24s and available for use by the M25s—which are, of course, immune to the disease."

VAATU-9 stood, and, glaring at the speaker, said, "I don't think you appreciate how many problems would be produced by this proposal. How would you get rid of seven billion M24 carcasses? Do you expect the incoming M25, as its first assignment on the planet, to bury all these bodies? Perhaps you expect maybe the on-site earth-monitors to volunteer to help bury them?

"Consecutive terminations in the past have always been gradual, taking sometimes millions of earth-years for resolution. Carcasses had time to rot and fossilize. Bones were covered with shifting sand and dirt. There was no infrastructure to preserve back then. All of this is different today. The M24 has produced a valuable infrastructure of buildings, factories, and roads, and maintains fields of cattle and other domestic animals it uses for its food. How do you expect to preserve this for the incoming M25?"

VAATU-9 turned his attention to the assembly at large. "Before you continue with your proposals for consecutive extinction, all of you must address the questions of how to dispose of seven billion dead M24 carcasses, how to maintain the domestic animal population—in fact, how to maintain every facet of the M24 infrastructure until the M25 arrives on scene. Finally, you all must consider how to introduce enough M25s into the system quickly enough to prevent widespread decay of the M24 world order. Otherwise, we will drop any further consideration of consecutive extinction and concentrate on concurrent termination techniques."

Another earth-monitor commented, "I think we can surely expect that at least one-third of the M24 bodies would have been buried or otherwise disposed of by surviving M24s by the time the last of them expired."

"I'm sorry, but I think you're being very optimistic," replied VAATU-9.

Hoping to bring discussion of Method A to a close, one of the Committeemen asked, "So are we saying that Method A should remain a consideration? Should efforts go forward to confirm and validate claims that a virus can be perfected that will terminate soon enough to prevent M24 countermeasures from succeeding?"

There was a buzzing in the crowd as they talked with each other.

Another Committeeman chimed in, "What do you all think? Can we see a show of hands of those suggesting further research into such a microbe?" Most of those present raised their hands.

VAATU-9 raised his hand to quiet the discussion, "One question before we vote. Can we stipulate that the experimental microbe we select will ensure a painless, relatively comfortable death for the M24? After all, it is the M24s that have brought our experiment forward to the incredible success it is today. Shouldn't we show some compassion to them before their souls are returned to our database?"

"Good point, sir," said another of the Committeemen. "I think this has to be considered. We all have to be grateful to the M24 and its designers for bringing our experiment to the level it is today."

There was a short delay while the Committeemen consulted among themselves. Then the meeting resumed. "The Committee has agreed to refer this question to microbe research for further review. We look at this method as possibly viable, pending further study," said the Chairman.

"Now, if there are no more comments regarding Method A, let's move on to our discussion of Method B."

Method B: Catastrophic Geographic Events appeared on the large screen behind the committee members.

"Although this method," said the Chairman, "has been effective in the past, we are not seriously considering this method, for reasons noted in the report. Are there any comments from the floor?"

Several hands were raised; one was picked. "Might not a meteor impact be one way to introduce the virus being considered for Method A into the M24 world? Perhaps if the meteor were to strike a less populated, more backward part of the planet, two objectives would be accomplished: the virus would be introduced, and significant portions of the poorly constructed infrastructure, portions that the M25 will certainly clear away in any case, would be eradicated. This would allow the M25 to start over in these areas with a clean slate, as it were. As you know, huge reserves of undiscovered mineral resources lie just under the surface in many of these backward areas. And we know that the M24 variants populating these areas have been a hindrance to the other M24 variants in their quest for resources."

Another hand rose. "We have to be careful with meteor impact insertion techniques. Let's be honest with ourselves. Meteor impact technology is very imprecise. If the meteor is too intense, reverberations will be felt around the world. There will be dangerous tsunamis, earthquakes, intense weather, and climate changes. Damage to valuable infrastructure around the world, not just locally, could be huge. The new M25 might face huge restructuring challenges. Conversely, if the meteor were too timid, our objectives would not be met; the M24 may survive long enough to fight and conquer the virus we inject. How can we be sure the meteor is sized and aimed to meet our objectives?"

"Point well taken. Let's hear more comment."

"Perhaps if the M25 inherited damaged infrastructure, it would feel spurred to make huge achievements," said an earth-monitor. "After all, we know that major calamities such as the US Civil War, the Great Chicago Fire, and the World Wars spurred the M24 to great advances in their civilization. We can expect from the M25 an even grander response to the challenge of damaged infrastructure."

"These are all good comments. Any others?" asked a Committeeman.

"The point was made earlier that we be compassionate toward our M24. How can meteor impacts and the resulting chaos be considered compassionate?"

"The impact point for the meteor was assumed, in the preceding comment, to be an area inhabited by less productive M24s, many of whom have held back the progress of the more productive populations. Ultimately, we have to face the fact that none of this undertaking can really be considered 'compassionate' in the full sense of the word. We just have to understand there are various levels of compassion. Some of the M24s will be treated with more compassion than others no matter how we end their time on the planet. In any event, their eleven-dimensional souls will all return to our laboratory database one way or another for a final, compassionate judgment."

A Committeeman raised his hand to stop floor comment and said, "All of your comments are superlative and are adding immensely to our considerations. We will certainly be looking into how meteor impacts can be calibrated and fine-tuned to provide exactly the effect we would desire for the M25 insertion. I suspect that if we were to err, it would be on the side of too much care rather than too little. In any case, you can all rest assured we will be seriously considering your comments."

The Chairman stood and said, "Now let's move on to a discussion of Method C."

The large overhead screen changed to show this new title: **Method C: X-rays or other appropriate cosmic rays to field-mutate the M24 DNA, eliminating or vastly reducing M24 procreation.**

"Let me remind you, before you comment, that controlled modification, or mutation as it's sometimes called, of the DNA of a living creature has not yet been successfully performed anywhere except under scrupulous conditions in the laboratory. The few field mutations performed up to now have unfortunately resulted in massive, unpredictable, and always monstrous change of the species for the worse.

"With that said, let me open the floor to comments. Any comments regarding Method C from you folks?"

A hand on the left rose first. "If we are successful in developing a precision DNA-mutation process for the field, won't there be mass agitation among the M24 when it discovers its devastating affliction? Particularly when it comes to realize the likelihood of extinction? And then as populations dwindle, its culture will unravel, wars will break out between the haves and have-nots, whole systems of service to the populations will break down. The endgame will be horrendous—probably much worse than any calamity ever experienced during and after its wars. Do we want this for our most successful creature design to date?"

A Committeeman stood and said, "Your thoughts are well taken. We need to consider how to reduce the awfulness of the end game for the M24."

Another earth-monitor raised his hand and asked, "Could M24 DNA be field-mutated so that it would birth only M25 creatures?"

Obviously annoyed that a question he thought had been answered was raised again, the Chairman decided it would be wise to expand on the subject. "In the past, we have looked into altering Method C in such a way that rather than reducing the M24 reproduction rate, we upgrade its DNA into M25 DNA by means of a very discrete cosmic particle bombardment. This is the ideal we have worked toward since the experiment began.

"Unfortunately, it's been our experience time after time that mutations of creatures in the field always produce damaged goods, not the substantially improved product designers wish for. We've tried many times to field-mutate lower level DNA code upward, but we have never been successful. So we are left, as we have always been, with the technique we know works: alter DNA in the laboratory and then physically implant the resulting experimental creatures onto areas of planet Earth where they can be expected to thrive. It is a tried and proven technique and still the only one we know to work reliably and consistently."

An earth-monitor, seeing the discussion was getting a little out of hand, stood up and asked a question to change the subject. "How sure are we that the M25 will provide a better, more humane culture on earth?"

A very exasperated Committeeman to the far left of the group answered this question. "Where were you when the meeting began? As our quality-assurance director very clearly explained to you all, there have been literally thousands of tests and retests of the various M25 characteristics, and revisions made to its DNA molecule ladder arrangement until all the design deficiencies we can find have been expunged. The M25 is as free of defects as our latest scientific test techniques can make it. The true test, of course, is in the actual deployment of the creature and for you earth-monitors to follow its progress. Your observations and reports back to us will be exceedingly important to us."

The earth-monitor frowned. "Is there any reason for us to worry that the M25 will in fact be a step backward, and that the civilization we've produced up to now will backslide toward the prehistory we saw for so long? Or that it won't progress— just remain status quo? Aren't we taking a big chance with this new species?"

Another Committeeman curtly replied, "Our tests tell us we are on the cusp of a major breakthrough. We don't believe we have anything to fear."

The earth-monitor sat down, satisfied that Committee predictions for mission success were on record and clearly defined.

At this point the Chairman rose and said, "Now if there are no more questions regarding Method C, in the interest of limiting this meeting, let's move on to a consideration of Method D."

Method D: Extinction by replacement life form
appeared on the large screen.

"We don't think much needs to be said about this form of mass extinction, but we will entertain questions from the floor."

No hands were raised.

The Chairman went on, "We don't see the M25 in any mind to exterminate *any* of earth's creatures, much less the M24, a creature we hope it learns to husband rather than exterminate."

"If there are no comments, let's move on to Method E."

Method E: Catastrophic climate change appeared on the screen.

"As you all know, this technique worked exceedingly well when we used it to terminate many of the earlier bipedal models," said the Chairman. "We knew none of them had the wherewithal to cope with modest changes in their environment and that they would succumb within centuries after the introduction of either cold or hot climate changes.

"But for climate change to kill the technically advanced M24, we would literally have to return Earth's climate to that of the Cryogenian period some seven hundred million earth-years ago, when the planet was entirely coated with ice and snow to a depth of several miles. And even then we think the M24 could evolve techniques to allow it to survive. To be absolutely certain of M24 extinction by means of climate change, we might have to return to the Hadean period, some four billion earth-years ago, when Earth was covered with masses of active volcanoes and climate temperatures were near the boiling point of water. Of course, along with the extinction of the M24 would be the complete destruction of the M24 culture and infrastructure, and we would be starting over from scratch. Obviously, this is not what we want. We want to build on the progress made by our M24, not completely start over."

An earth-monitor several rows back raised his hand. "Question: Could we not think in terms of a relatively mild climate change, with a pollutant in the air that attacks the lungs of the M24 and produces cancers to the point of extinction? The climate then, without the M24 present to contaminate it, would in time revert back to the purity it once had before we introduced our bipedal hominid."

An engineer on the Committee replied, "We think that such a pollutant would also kill off vast portions of the vertebrate animal world as well, thus making the world in which the M25 finds itself very difficult for it to thrive. It would be a big setback for our experiment. Are there any more comments about the pros and cons of Method E?"

Silence for a few seconds. Then the questions, suggestions, and discussion of Method E continued for another half hour, until VAATU-9 stood up, raised his hand for silence, and thanked the assembled earth-monitors for their comments and input.

"All of your comments have been recorded and will be seriously considered before a decision is made regarding a final solution of the M24 question. Now, before we leave, are there any questions of a general nature?"

CHAPTER 13

COMPASSION

"I hate to bring up this question again, sir," said someone in the third row, "but many of us earth-monitors have, in the course of our lives on Earth, developed rather close relationships with the M24s. Some of us have even married them," he said, looking around the room, uncomfortable having to ask the question. "Of course, we are unable to procreate with them, but nevertheless some have offspring from previous unions. What is to become of the offspring when this M24 extermination plan is implemented?"

"Hold on one minute, sir! This is not an *extermination* plan," said VAATU-9, pointing directly at the man. "It's an *extinction* plan. Remember that. There is a big difference."

"Of course, sir."

"We are not Nazis. Our experimental creatures, including the upright bipedal creatures, become extinct. They are not exterminated. Is that clear?"

The dead silence in the room suggested he was clearly understood.

"Now to get back to your question. We are glad you asked this question and are happy to expand on what was said earlier. The Committee has worked out protocols for both consecutive and concurrent earth-monitor family extinction plans. If the Committee, after deliberation, selects a consecutive extinction plan for the M24, all earth-monitor M24 family members will pass in accordance with plan requirements, because the plan requires the planet be completely clear of M24 before introducing the new hominid. No exceptions." VAATU-9 looked around the room for emphasis.

"If the Committee concludes that a concurrent extinction plan for the M24 best meets project objectives, your family M24s, age forty and older, will get immunity and will be allowed to live out their natural lives. Earth-monitor family M24s aged thirty-nine and under will be subject to the requirements of the extinction plan we select. This includes the M24 children and friends of any earth-monitor family member. Again, no exceptions."

VAATU-9 waited for the earth-monitors to digest what he said, then he motioned to another Committee member for more comment. The Committee member stood and added, "We know that you, as members of our experimental team, will understand the necessity of this rather severe circumstance and will adhere to our plan. Details and an opportunity to ask Committee members specific questions will be provided at the end of the meeting."

"Any other comments of a general nature?" asked the Committee Chairman.

After a moment's silence, comment buzzed around the room. Finally, one earth-monitor spoke up. "If, in a concurrent plan, life for earth-monitor family M24s aged forty and above becomes unbearable, what would be the plan to keep them alive?"

"They would be brought aboard one of our mother ships for transit to a place where they can be cared for. They would carry on a decent life, perhaps not quite what they had planned for themselves, but nevertheless a healthy and happy one. In some cases we think their lives would be an improvement over what was available to them on Earth."

Several in the audience expressed discomfort with this last comment. They had strong doubts about the feasibility of the plan—it wasn't at all what they expected. They had thought the extinction process would extend over an earth-century, allowing all M24s to die of natural causes.

After more discussion it was clear the Committee wanted to complete the replacement of the M24 in record time—the previous twenty-three hominid extinctions had extended over dozens of earth-centuries. This time extinction would occur in a matter of earth-years. No doubt the Committee was very uncomfortable with the 20 percent M24 defect rate revealed in the *Visitor's Report*.

The rest of the meeting passed in a daze of discussion and argument, none of it meaning much. When it was over, Committee members spread out to answer individual questions and amplify what had been said in the meeting.

VAATU-9 approached one of the individuals near me who had sharply questioned the proposals to end the M24 time on Earth. VAATU-9 was tall, athletically built, and larger than most others of the Committeemen. He had a commanding presence, with his large, black, hypnotic eyes dominating a somewhat-larger-than-usual head. His was a friendly face, nevertheless, and it projected, along with his firm handshake—usually involving both hands—a heartfelt feeling of interest and respect in all whom he met.

"Hello, sir, what did you think of our presentation?"

"May I speak frankly, sir?"

VAATU-9 nodded. "Of course."

"In all honesty, I think you are all monsters and totally uninterested in the fate of mankind. How could you think this way?"

I could not hide an involuntary wince. This was not going well at all.

"Whoa there. Back off. We're just as committed to the health and happiness of mankind as you are. We really are! But apparently there is an important difference between you and me in the way we approach our commitment," said VAATU-9. "We cherish the M24 and all the inhabitants of planet Earth, maybe not with pure love, but certainly in the way the General Motors design engineers cherish each of their automobile products—with great affection. And when our 'product' ends its life, unlike automobiles, which go to the trash heaps around the planet, we return the 'product's' Basic Input/Output Modulator—its 'soul,' as the M24 likes to call it—to a very pleasant existence in our laboratory database for judgment and relocation. The eighty percent who have led lives we judge to be productive and acceptable remain there free to mingle with all their associates who have preceded them home. They blissfully resume their associations, relationships, and friendships as they choose."

The earth-monitor tried but couldn't get a word into the conversation.

"As you know, about twenty percent of the M24 are judged to have fundamental defects. After all, Hitler didn't round up six million Jews and incinerate them all by himself. These defectives, of course, are deleted from the database so as not to contaminate the non-deficient, or, heaven forbid, be reused."

VAATU-9 continued, "Ask yourself, sir, how long would the General Motors Corporation make cars if twenty percent of them were defective? I'm afraid you don't look at mankind, as you describe the M24 civilization, the way it is in actual reality. Today, in the M24 world, cell phone designers, automobile manufacturers, and designers of any other product have no compunction whatsoever to ending the 'life' of an obsolete product in favor of bringing in an improved model. When one of the large M24 corporations or businesses becomes ineffective or unable to compete in the marketplace, it collapses, and hundreds if not thousands of humanoids are cruelly put out of work."

VAATU-9, waiting a moment to let that sink in, added, "In exactly the same way, if we see an opportunity to improve our 'product,' we take it. Sadly for the M24, we are at that point in our experiment when a replacement product is needed."

"You have a very cruel outlook," the man shot back. "The M24 is a living, breathing human, and unlike a cell phone he has feelings and compassion. Is there any compassion in your equations?"

"Assuredly there is. Just as these people try their best to assist persons put in terrible situations by recessions and disease, we are, as you heard in our discussion, looking for the least painful means to delete the M24 from the planet. A final decision hasn't been made. But when it is, you can be sure it will reflect our appreciation for all the successes the M24 has provided us. If we determine that a pandemic will be the way for the M24 to go, the disease will be fast acting and as painless as possible. As a matter of fact, the death we plan for the M24 will, in most cases, be far more compassionate than the fate that usually befalls them: the violence of automobile accidents; the agony of cancer; the profound, wrenching pain of heart attack; the extended, slow, painful disintegration from Lou Gehrig's disease. The list of bad ends for the M24 is a long one.

"Every one of the M24s knew when they were born they were going to die. I'm sure you remember the M24 comedian who once said being born was bad for your health. This way the M24 will depart planet Earth in the most compassionate way we can devise.

The earth-monitor was not sold. "You are utterly cruel. You are without any compassion whatsoever."

"I'm sorry you feel this way, sir. But aren't the M24s cruel in the same way to their laboratory test animals used in research for new medicines? Many of them put their pet dogs and cats 'to sleep,' don't they? Is there any difference? And you're well aware, I'm sure, that we have, shall we say, 'put to sleep' all the previous twenty-three humanoid models, not to mention millions of other lesser species."

An earth-monitor approached as they talked and motioned to VAATU-9 that it was getting late.

"I really must head back to the plane now. It was a pleasure talking with you," VAATU-9 said warmly as he headed to the door.

VAATU-9's cheery comment did nothing to lessen the churning in the stomach of the man who had so thoroughly criticized VAATU-9, and some of the other earth-monitors as well who had never been through a hominid extinction before. And this was going to be a bad one.

CHAPTER 14

EXTINCTION

Four months after the momentous Roswell meeting of earth-monitors, the Committee verified that indeed it was not only possible but certain that Method A microbes could be developed and introduced to Earth effectively and efficiently to produce a massive, relentless pandemic, unheard of since smallpox exterminated entire tribes of Native Americans after the arrival of Europeans in 1492.

This time, victims would first notice lethargy, an exhausting fatigue, and an overwhelming desire just to nap. Very soon, naps would become restful sleep, and attempts to waken the patient would be difficult. By the time a doctor was summoned, the outcome was hopeless. To satisfy friends and relatives, the doctor would draw blood samples and find there in the clear glass tube of his syringe exactly what he expected: a pinkish, creamy fluid, flecked with small, white coagulates instead of the bright red of healthy blood. No need for laboratory tests. The prognosis: a virulent, terminal form of aggressive NK-cell leukemia—what soon would become known to Earth's population by way of the sensational tabloids as "galloping leukemia." The patient would have only days to settle his affairs and prepare for the end. It would be a peaceful, painless end, but also, of course, devastating to friends and relatives.

Committee members had argued over a period of many weeks after Roswell, in meeting after meeting, about the most efficient yet compassionate means of putting down the M24. The old favorites—bubonic plague, smallpox, and botulism—were of no use, either because the M24 had found remedies or because these afflictions would be too painful and traumatic. No, only a painless, fast-acting disease, which the M24 could not possible defeat, would be suitable for ending the M24's involvement in the ongoing experiment. After all, Committeemen have feelings too. They had grown to love each of the life forms they had watched thrive on their lovely planet, and it was important to them that each of their humanoid creations receive credit and appreciation for its part in advancing their science.

What's more, Committeemen knew their hardworking earth-monitors had grown attached to their subjects and would expect the end of the M24 to be respectful and kind. And they knew that as the M24 expired over a period of years, these earth-monitors, saddened to have to stand by powerlessly as their M24s passed away, must at the same time busy themselves tirelessly guiding and settling the arriving M25s into their new communities around the globe. It would not be an easy time for the earth-monitors at all.

After studying and reviewing proposals for the introduction of numerous different microbes into the M24 population, the Committee narrowed the number of proposals for consideration down to seven, then to four, and finally to two candidates, which tests showed would have no more than a 0.02 percent chance of defeat by the M24 medical profession. One proposal was for an advanced, fast-acting form of the Spanish flu that had killed fifty million M24s after their World War I. The other was a microbe-produced variation of leukemia. Similar in nature to forms of a cancer the M24 had so far been unable to overcome, it would kill its victims in less than two weeks.

After many slide presentations by laboratory personnel, and weeks of exhaustive Committee meetings evaluating each proposal in detail, the decision finally came down in favor of the leukemia microbe. Those favoring the flu microbe could not, in the end, convince Committeemen that the disease was completely free of victim distress, whereas the leukemia approach, every bit as lethal, just put its victims to sleep.

During the Committee meeting at which approval was given to the leukemia approach, one technician summarized the feelings of many in the room when he said, "The M24 should have no difficulty understanding that a fast-acting leukemia is the best possible means to end their tenure on Earth. Surely they'll understand. After all, they commonly put their dogs and cats to sleep."

Since Method C, the introduction of carefully selected cosmic rays to impair sexual reproduction, had also been committed to, it was determined to proceed with this method as well as Method A. The first signs of a devastating cancer pandemic would soon appear and take up all the attention of M24 medical professionals. They would never notice the subtle onslaught of cosmic particles attacking human replication.

* * * *

The earth-monitors could only watch as the M24 population slowly flickered out of existence. As its agony mounted, the stricken *Homo sapiens sapiens* looked back at World Wars I and II and the devastation visited by Islamic terrorists on Middle East populations in the early years of the twenty-first century as insignificant by comparison.

The first of the pandemics hit the world's population in October, 2033. Medical facilities around the world united in desperate research to vanquish the disease, but they could not stop its spread before nearly two billion fell victim. If there was any comfort to the surviving loved ones of the departed, it was that people in every village, town, and city in the world were experiencing similar grief.

Just as bubonic plague had done in the sixth, seventh, and fourteenth centuries, whole cities were cleared of human life. Governments failed their constituents due to lack of revenues and personnel, and military organizations disintegrated. Soon the use of suicide bombs in the Middle East ceased as the most fundamentalist of believers waited patiently for their journeys to eternal Bacchanalia.

Underreported by a news media struggling with a lack of personnel was the mysterious appearance of people in or near the devastated areas who were somehow "different." Uncounted thousands of these people and their families, seemingly immune to the killer disease, appeared in towns and skillfully replaced many of the key people in government, industry, and education who had perished. Soon, the chaos diminished thanks to these unusually gifted and adept "replacements," and life began to return to some form of normalcy.

Two more devastating worldwide pandemics followed in 2039 and again in 2046, and by rough count, two billion more M24 perished in each. Small island nations and tribes in remote corners of the earth were entirely wiped out; a rare few were repopulated with "new people."

Economies imploded after the onslaught of each new pandemic, governments failed to provide many of the required services (tax collection not being one of them), and armed forces retrenched and maintained order as best they could. The medical profession collapsed and did not recover until it was able to bring in replacements, who seemed to adapt quickly to their duties.

The pandemics rampaged on unabated, unaffected by any treatment that could be devised, until mysteriously petering out as pandemics always had in the distant past. By the year 2050, the M24 population had shrunk to less than a billion and was almost totally sterile. The "new people," now numbering many billions, were assuming control and responsibility for the well-being of nations around the world and were doing rather well at it.

The world's governments had long since recognized UFOs as, in fact, extragalactic visitors and final proof of an omnipotent but somewhat error-prone Creator. The Age of Darwin was over. All of Earth's political entities and religions were in shock and confusion, and, in 2057, when computer hackers finally got their hands on a recent edition of *The Visitor's Report* and gleefully released it to the public, any hope for survival the vastly depleted M24 population may have harbored vanished entirely. As they approached their demise, the remaining M24s now knew the despair and anguish felt by the previous twenty-three humanoid models. They no longer mattered, and they knew it. The last *Homo sapiens sapiens* on Earth passed away in 2089, at the age of 103. No one cared.

PART III

CHAPTER 15

Tranquility

Sixty earth-years after the momentous Roswell meeting of 2030, earth-monitors looking after the affairs of the M25 do so with very mixed feelings. It is, of course, with profound relief to everyone that the turmoil and vast disruptions caused by the reduction of the obsolete humanoid model are over. World affairs are conducted now by the new M25, and they proceed smoothly and with no sign of significant disorder or antagonism.

The Committee is delighted to see it achieved its prime project objective: the abolition of crime and violence, which has almost disappeared, to the relief of nearly all honest citizens of the world. The world's military organizations are 10 percent the size they were when the M24 ruled Earth, and they are still contracting in size and lethality. The M25 world community is working smoothly; earth-monitors have little to do and little to report to their vehicles patrolling, in much fewer numbers, the heavens above them. Earth-monitors see this as a sign that their own future may be limited, and they are expecting their numbers to diminish.

The M25 has turned out to be an individual who lives in a comfort zone: a cozy home inhabited by a husband, loving wife, and three or four kids (preferably three). More than four kids seems excessive to the intelligent, sensible M25 family. Three kids bring balance to the family. Three allow time for sufficient attention to each child while at the same time allowing more time for enjoyment and vacations away from the kids. More than four does not.

A forty-hour workweek seems sensible to the well-adjusted M25, a thirty-six-hour week even more so, and so does saving money for a comfortable retirement. The only major and necessary expense needed is money to put their brilliant children through the six to eight years of college usually required to pursue professional careers. Once out of college, the kids are out of the house and on their own. What is the point of working for millions or billions of dollars that one can't spend in a dozen lifetimes, funds that would only spoil the kids?

Societal violence is pretty much limited to TV reruns of old shows produced for the M24 a century before. Children are interested in these violent shows, but most parents sensibly refuse to allow their kids to see them. After all, violence is frowned on as a way for intelligent adults to settle their arguments.

Not long after the M25 became a majority of Earth's population, earth-monitors noted an upsurge in the popularity of chess. Chess leagues formed, TV shows more and more chess matches, and chess commentators are well known and rated by fans. The best chess experts are written about on blogs and are minor celebrities to the many adherents. Some actually have huge followings of teenagers and lonely women.

The M25 enjoys watching sports, but does not have the zest for adventure nor the aggressive nature needed to participate in professional football and ice hockey. Why, they sensibly ask, recklessly endanger their bodies just to move an oddly shaped, inflated ball back and forth on a grass field lined every five yards with white lines? Why put on ice skates and race back and forth on an ice rink trying to shoot a small round piece of hard rubber into a space enclosed on three sides by netting? It all just seems silly to the passive M25.

Tennis is seen as a silly sport, perhaps only good for children while building their bodies. But just hitting a ball back and forth across a net is an exercise in futility to the highly intelligent adult M25.

Golf is now seen for what it really is: addressing a small white ball, hitting it with a special club as far as one is able, then chasing after it in a battery-operated four-wheel cart, only to swat it again and again until finally urging it into a small round hole in the ground. And repeating this process for seventeen more trials before ending the day in a nearby bar or pub. Any sensible person would ask: Why not pass up the inevitable frustration of the golf course and go directly to the bar?

During the transition years, life for the earth-monitors was pure hell. We manned the transit stations as millions of M24s perished and other millions of M25s arrived and received directions to their various locations and assignments. Masses of documents hidden in computer storage units around the world had to be located and surreptitiously revised to reflect the loss of millions of M24s and their replacements by arriving M25s. Earth-monitors hacked into computers and retrieved and altered documents to allow M25s to blend into their new communities around the world. Earth-monitors were pretty good at this, but the few inevitable screw-ups caused immense confusion and lots of time to correct.

As M24 losses mounted, the burial of the dead and care of the survivors was almost more than the earth-monitors could bear. Earth-monitors were forbidden to identify themselves as such to the M24, but they could pitch in and help with what little they could provide. Tens of thousands of new earth-monitors were brought in temporarily to assist with relocation, documentation, and finding vacated homes for the new people to move into. The arriving M25s were told only that people were there to locate them in their new assignments and give advice.

It was an exhausting time for all earth-monitors, especially those who lost loved ones during the leukemia pandemics, but somehow we got through it all.

CHAPTER 16

Aftermath

September 28, 2090, is a rainy, dreary day in Port Saint Lucie, Florida—a perfect time and place for my chronicle to end, and an appropriate place to insert a letter I bravely wrote. Well, maybe not so bravely after all, written as it was just weeks before my official retirement.

VAATU-9
(Address deleted)
   Content with the life its intelligent population provides, the M25 has become complacent, almost entirely uninterested in improving anything about its serene life. The M25 is perfectly happy with the vehicles it drives on the highway and in the air. It sees no need to improve factory processes, construction techniques, or communication devices. Scientific progress has come to a halt. Everything works just fine.

Life has arrived at perfection. How, the M25 asks, can one possibly improve on perfection?

And this is a problem. For many years after the M24 extinction, earth-monitors have been disturbed by a strange, unexpected turn of events—one that has dire implications for the M25, a consequence entirely unanticipated by the Committee. The combination of diminished aggressive nature and increased common sense, along with high intelligence levels in so many of the M25, has produced an individual with little incentive to improve his situation.

The M25 hasn't the incentive or motivation to search for better medicines and techniques to care for the sick. Most of them are healthy due to their natural resistance to cancers and other diseases and injuries, and their bodies have the ability to rejuvenate serious brain and nerve injury, making long painful rehabilitations a thing of the past. As a result of the revisions made to their DNA, the few who incur really serious injury rarely experience severe pain and, in any case, are usually too far gone to be saved. So medical problems seldom motivate the M25s to take on the risks and heavy expense involved in developing new medicines for disease prevention and improved techniques to heal injuries.

Earth-monitors report their observations of this serene, sensible, and very contented population to the Committee in their periodic reports. One such report described the M25s as behaving like herds of sheep. Another described them as Stepford wives—*and husbands*—an allusion to a 1975 Hollywood movie involving "ideal" robotic wives. To the earth-monitors, experienced as they are reporting on world wars, pandemics, and violent crime, life in general now is boring to them—yes, boring. But not to the sensible, highly intelligent, tranquil M25.

Most problematic of all is the refusal of the M25 to recognize or be concerned by the planet's disappearing underground energy resources. Committee members were sure the superior intelligence bestowed on the M25 would spur it to develop simple, low cost, radiation-free fusion energy[21] to replace the widely used fission-derived energy process by which tons of highly radioactive waste byproducts are produced annually. This would limit the need for oil and gas energy, and almost entirely end the use of coal. But this has not happened, and it is becoming clear that it will never happen as long as the M25 is so content with its state of being.

It is dawning on more and more of us earth-monitors that by reducing aggressiveness and increasing intelligence and good judgment, the Committee engineered out of the M25 DNA the very characteristics that impelled its predecessor to rise up out of its primitive world and civilize itself. Many earth-monitors believe the Committee should have known that highly intelligent people with a limited aggressive nature, living an entirely satisfying life, seldom stick their necks out looking for radical new ways to change the world around them. Lord knows there were plenty of examples of this in the *Homo sapiens sapiens* world.

You need only to refer back to your old *Visitor's Report* from 2030, to the part describing the changes that were to be made in the creature's aggressive nature. Sure enough, there for all to see were sections 5D and 5E, calling for higher intelligence and less aggression. Everyone saw it. It was there in plain sight. And no

---

21 Current efforts to develop fusion nuclear energy are only half-hearted and far too naïve, complicated, and expensive to be practical. They are going nowhere.

one recognized that here was a prescription for a lazy man.

The M25 is safe from the aggressive behavior of others, certainly. But also quite satisfied with his station in life, and, most importantly, absolutely sensible in everything he does. The Committee did realize that the champions of *Homo sapiens sapiens* industry, authors of the best literature and music, the best athletes, and the very best inventors and seekers of new and better ways of living were all intelligent, but the Committee failed to note they all had very aggressive natures. They were obstinate, pushy, driven people who seldom took no for an answer. Whether or not they were liked, they were all respected. None of them were ever thought of as sensible.

If the Committee members had simply done more homework, they would have seen the M24 leaders of change in their world were quite dissatisfied with what they saw around them, often led disastrous lives, and aggressively worked outside the mainstream to change their world. Henry Ford was stubborn, intractable, and quite abusive to his son Edsel. Some believe this led to Edsel's early death. Charles Goodyear, in and out of debtor's prison, struggled to find a way to vulcanize rubber, only to be cheated out of any profit from his inventions. Goodyear died alone and deep in debt. The Wright brothers never married; instead they obsessively devoted their lives to demonstrating manned flight and then to protecting—unsuccessfully, I might add—their precious patents. Steve Jobs and most other world changers led unusual lives, were considered odd during parts or even most of their lives, and pursued goals the average citizen did not understand or really care about until after the goal was reached.

Years before, when the M25 DNA molecule was under development in the laboratory, the Committee engineers seemed determined mostly on finding and deleting the groupings or combinations of DNA molecules (genes) that caused evil intent in their creature. Their prime objective was to prevent the emergence of any more Hitlers, Stalins, or Maos, who had caused such extraordinary mayhem. Other creature alterations or improvements were secondary. At no time did they recognize or appreciate the *value* of the M24's aggressive nature, the good side of it.

Committeemen all know the prime objective for the M25 on Earth is to bring in new ideas and products for the betterment of its population. But *it* doesn't know any of this. If the M25 doesn't produce, sooner or later the Committee will be forced to develop an M26, an M27, an M28—as many new creature designs as it takes until the molecular combinations that drive their creature to improve its world are isolated from those that cause evil.

Let's face it. It took twenty-four tries before the Committee found a creature that could civilize itself. They will not hesitate to keep developing newer and newer humanoid models until they get it right. Worst case: bring back the M24 and resume the appalling violence that will assuredly accompany the scientific advances they produce—a Hobson's choice if there ever was one.

Respectfully submitted,

* * * *

As I end this document, I can report the Committee is at last acutely aware of the M25 motivation problem, and a new hominid model, the M26, is indeed in the works. Recent laboratory tests uncovered deficiencies in early experimental versions of the unreleased M26, so a new, further improved hominid laboratory model, currently designated the XM26-4, has been designed and is undergoing final acceptance tests.

A mass extinction plan, which worked so well removing the M24, is still "on the table," so to speak, for the M25 if it remains unable or unwilling to bring forth new improvements for society—particularly safe, clean, radiation-free fusion nuclear power for the planet's future energy needs. Dependence on chemical fuels and on "dangerous" nuclear fission, which continues to produce vast storage fields of spent radioactive fuel rods, is not an option.

The M25 has the smarts, but so far there is no sign it has the drive, willpower, and willingness to take on the enormous effort involved in producing fundamental change for its society. As things stand in 2090, the M25 seems no more likely to make notable change to its society than did *Homo neanderthalensis* in the three hundred thousand years it inhabited Earth or Cro-Magnon man in its twenty-five thousand years on earth. During their long tenures on Earth these creatures could not figure out how to mine coal, drill for oil, smelt copper, harness farm animals to do their work—or even invent the wheel.

The Committee had taken just fifteen thousand earth-years to design, test, implant, observe, and then exterminate the M24. With its modern advanced laboratory tools and precision database of information, it shouldn't take the Committee anywhere near that long to arrive at a decision for the M25. Earth's energy needs are increasing exponentially—as is its population. A mass extinction plan for the hapless M25 is way overdue. Tick tock. Tick tock.

AUTHOR'S NOTE

The Visitor's Report is intended to provoke thought and, I hope, discussion among you and your friends about reality, life, and humankind. It is nothing more than a story about the universe coming at you from a different point of view, one I find difficult to argue against.

Maybe it was my Aunt Bee who started me thinking "differently." When I started to look into the anthropology of ancient man, it was not long before my childhood memories came back to me of Aunt Bee: dotty, but lots smarter than I, and particularly skilled in the art of solving giant, insanely difficult jigsaw puzzles. She loved the challenge, I guess, and was so good at it that she eventually attacked only the very largest and most complex puzzles, especially those without a picture of the completed puzzle on the cover of the box to provide a hint to guide her.

One day, when I was seven, I approached the card table on which she had begun her latest puzzle. About fifty pieces were on the table, and she was trying to arrange them in such a way as to suggest what the picture would be. I looked at the pieces for several minutes and abruptly said, "You know, Aunt Bee, I think the picture will be a tree. Look…those pieces could be a trunk, and those could be branches. And look. There. If you turn that piece a certain way, it could be a branch leading away from the trunk."

I could see she was cross when I said it. Without looking up from her puzzle, she murmured, "Bobby"—I hated that name—"aren't you a little young to tell me how to do these puzzles? There aren't enough pieces on the table yet to tell you what the picture is."

I returned an hour later. Aunt Bee was still hard at work, but she had changed her tune. "Bobby, you're right about that tree. I'm sorry I jumped on you. This puzzle *is* going to be a tree." I was quite pleased with myself.

Days later, there was Aunt Bee with more pieces of the puzzle in place, furiously rearranging them, first in one tree arrangement, then, as more pieces were added, in another.

"Bobby, you know, I think you're right about the picture being a tree, but now it looks more like a bush. There are too many offshoots and branches going in every direction to be a tree. Every time I add a piece to the puzzle, I have to rearrange the whole pattern to make a tree. I'm going nuts."

Aunt Bee was no longer herself. She wouldn't eat; she couldn't sleep. None of us could pry her away from her puzzle. Anytime Mom came by her table and suggested maybe it wasn't a tree, Aunt Bee flew into a rage. As the days passed she became still more obsessed with the puzzle. She was determined to form a tree if it was the last thing she ever did on this earth.

As it turned out, this puzzle and her search for a tree were indeed the last things she did on this earth. She collapsed at her table from a stroke while furiously rearranging the pieces into what she was sure was a bush. To the very end, she refused to listen to any of us who suggested the picture was not a tree. Weeks later, Mom finished the puzzle for her. It wasn't a tree—or even a bush. For years afterward I was sure I had killed Aunt Bee.

Now I compare paleoanthropologists to my Aunt Bee. They find ancient bits and pieces of fossils that someday will give us a picture of our heritage, and arrange them to form a tree. Today there are still so few fossils to work with that the picture is not in focus. But because someone in 1859 suggested the picture *will be* a tree, they furiously rearrange their fossil bones each time new ones are found until they again see the hint of a tree, or maybe a bush.

And, like my beloved Aunt Bee, paleoanthropologists get very cross if someone suggests the picture might not be a tree.

Appendix A

The Four Fundamental Forces of Nature

There are four fundamental forces of nature.

1. *Gravity* is the force that pulls us to the surface of the earth, keeps the planets in orbit around the sun, and causes the formation of planets, stars, and galaxies.
2. *Electromagnetism* is the force responsible for the way matter generates and responds to electricity and magnetism. It is used in nearly all our household appliances.
3. *The Strong Nuclear Force* acts only inside the nuclei of atoms and binds the tiny particles (quarks) together in an atom's nucleus.

4. *The Weak Nuclear Force* is responsible for certain kinds of radioactive decay, such as the decay measured by archaeologists when they perform radiocarbon dating.

All the forces except gravity are described using quantum theory, which states that these forces are transported by tiny particles.

Gravity is best described using Albert Einstein's Theory of General Relativity, which is not a quantum theory. Instead, it imagines that gravity is generated when matter distorts space, like a heavy object stretching a rubber sheet supporting it. Smaller nearby objects then "roll" downward, toward the larger object.

Einstein's general relativity predicts that the force of gravity increases to infinity in black holes—a nonsensical notion that means general relativity is flawed. Many scientists are working to formulate a quantum theory of gravity, in which the force of gravity is carried by small particles called "gravitons."

One approach to quantum gravity is known as *string theory*, which treats infinitesimal particles as if they were tiny one-dimensional strings of energy vibrating at various frequencies to differentiate themselves from each other. This theory requires the existence of multi-dimensional reality, anywhere from five to twenty-six dimensions, possibly more. String theory is expected to describe, in time, what occurs in the thousands of black holes discovered in the outer space of our galaxy.

APPENDIX B

SOME OF MANY PHYSICAL CONSTANTS REQUIRED FOR LIFE[22]

The controversial *Anthropic Principle* is a multi-faceted explanation of the many life-supporting "cosmic coincidences" supporting life in the universe. There are many long and very interesting discussions of this principle available from a variety of sources, too many to go into here. Listed here are just a few "cosmic coincidences," or physical constants, that are required for life to exist.

---

22   Adapted from *A Seminal Presentation by Astrophysicist Dr. Hugh Ross, given in South Barrington, Illinois, April 16, 1994* concerning 41 Fined-Tuned Characteristics for a planet to support life.

1. Electromagnetism

Unless the electromagnetism force takes on a very special value, molecules won't form. If the force is too weak, electrons won't orbit their nucleus; if too strong, it will prevent molecules from forming.

2. Strong Nuclear Force

This force holds protons and neutrons in the nucleus of an atom. If it is too strong, hydrogen, the basis of life, could not exist. If the force were weaker than it is, protons and neutrons would not form the nucleus of atoms; hydrogen would be the only atom formed.

3. Mass of the Proton and Neutron

The proton has slightly less mass than the neutron. If this ratio were to change only very slightly, either there would be too many protons or too many neutrons for life to exist.

4. Gravity

For life to exist in the universe, gravity must be 1×10^{39} times weaker than the force of electromagnetism. Yet planets, stars, and galaxies will not form unless gravity is dominant in the universe, so the other forces must cancel themselves out and leave gravity dominant.

5. Electrical Neutrality

The numbers of positively charged particles must be identical to the number of negatively charged particles; otherwise, electromagnetism will dominate gravity, and planets, stars, and galaxies will not form.

6. Electrons Must Equal Protons

The number of electrons must equal the number of protons to extremely close tolerances.

7. Mass of the Universe

If the mass of the universe were larger, the matter in the universe would be in the form of elements heavier than iron. If the universe were less massive, the only element produced would be hydrogen, and perhaps a small amount of helium.

8. Age of Star

Early formed stars do not have the essential heavy elements necessary for life. Only spiral galaxies form stars late enough to produce the heavy elements we all know are needed for life.

9. Position in the Galaxy

Life is impossible in the center of the galaxy and in its outer fringes. Life is only possible two-thirds of the distance from the center. The stars at the center are so close to each other that their mutual gravities would destroy planetary orbits; farther out, there are too few heavy elements from exploded supernovae available.

10. White Dwarf Binaries

Fluorine, one of life's essential elements, is produced only by white dwarf binaries, stars orbiting other burned out stars.

11. Properly Sized Sun

Stars larger than our sun burn too erratically and too quickly to permit life to form on planet Earth. A star any smaller than ours would not provide the heat needed to warm our planet and sustain life unless we were closer to the sun, which would introduce other problems for sustained life.

12. Tidal Forces

If our planet were any closer to our sun, tidal forces would increase by the inverse fourth power to the distance separating them, causing Earth's rotation period to lengthen. This happened to Mercury and Venus, whose rotation periods are now several months. On Earth this would cause days to be too intensely hot for life, and nights too intensely cold for life. As it is, Earth's period of rotation is slowing, but at an extremely slow rate.

13. Speed of Earth's Rotation

Early in Earth's history, Earth's rotation period was twenty hours, too fast for life to exist, because weather was far too violent. Someday this rotation period will be twenty-eight hours, too slow for life to exist. Temperatures during the day would climb to 170 degrees; at night the temperature would fall to minus one hundred.

14. The Planet Must Be Right-Sized

If the planet is too large, its mass will cause it to accumulate toxic gases, such as ammonia, methane, hydrogen, and helium, in the atmosphere. If too small, the planet will not retain enough water. Methane's molecular weight is 16. Ammonia's molecular weight is 17. The molecular weight of water is 18. The planet size must be fine-tuned to keep the water but reject the ammonia and methane.

15. Jupiter

A large planet such as Jupiter is needed in an outer orbit to help deflect incoming asteroids and planets that would otherwise collide with Earth.

16. The Moon

There must be a single moon to stabilize the rotation of Earth and hold it to a tilt in its axis of twenty-three and a half degrees, which produces seasons needed for life. Lacking a moon like Earth's, Mars's axis moves through a range of zero to sixty degrees, which makes life as seen on Earth impossible.

Appendix C

Taxonomy of Bipedal Creature Design Series

Long after the Committee began its three-dimensional experiment, data returning to them from their experimental planet Earth indicated that progress would not continue unless they designed an upright, bipedal creature equipped with an advanced control module (brain) and with its two front legs modified into five fingered hands and arms to enable it to more efficiently adapt to and affect its surroundings. Thus began a series of test creature design families.

Below is listed the Committee bipedal species design series, arranged by model number, introduction date relative to Earth-time, and creature identifications assigned by *Homo sapiens sapiens* paleoanthropologists.

M = Model number assigned to its experimental creatures by the Committee.

MYA = Million years ago

KYA = Thousand years ago

EMH = Early Modern Humans, i.e., prominent brows, receding chins, projecting face; powerful body; mix of archaic and modern traits; modern brain case.

AHS = Archaic *Homo sapien*; robust skeletons; prominent brow ridges; lived a physically challenging life.

AMH = Anatomically Modern Humans, i.e., prominent chin; vertical forehead; brow ridge barely visible; gracile body build, behavioral modernity.

BIPEDAL SPECIES LISTED BY COMMITTEE MODEL NUMBER

M	BIPEDAL SPECIES	MYA	DATE, PLACE DISCOVERED	COMMENTS
1	*Sahelanthropus tchadensis*	7	2001, Chad	Named "Toumai"
2	*Orrorin tugenensis*	6	2000, Kenya	Discovered by Pickford, Cheboi, et al.
3	*Australopithecus anamensis*	5.5	1967, Kenya	Small brain—400 cc
4	*Ardipithecus ramidus*	4.4	1994, Ethiopia	Discovered by Yohannes Haile Selassie
5	*Australopithecus afarensis*	3.8	1974, Tanzania	First discovered by Mary Leakey; 600 cc brain size
6	*Australopithecus bahrelghizali*	?	1995, Chad	Discovered by Michel Brunet

7	*Australopithecus africanus*	2.8	1947, South Africa	Discovered by Robert Broom
8	*Paranthropus aethiopicus*	2.5	1985, Kenya	"The Black Skull"
9	*Homo erectus*	2.04	1986, China	Wushan Man; 1,000 cc brain size
10	*Paranthropus robustus*	2	1938, South Africa	Discovered by Gerte Terblanche
11	*Homo habilis*	1.9	1973, Kenya	800 cc brain case; big increase in cognitive creativity; first to make stone tools.
12	*Australopithecus sediba*	1.95	2008, South Africa	Humanoid hands, ape-like feet.
13	*Parenthropus boisei*	1.8	1959, Tanzania	Discovered by Mary Leakey
14	*Homo georgicus*	1.8	2001, Georgia	4.5 ft. tall; small brain
15	*Homo ergaster*	1.75	1975, Kenya	Discovered by Bernard Ngeneo
16	*Homo gautengensis*	1.75	1976, South Africa	Discovered by A.R. Hughes
17	*Homo erectus*	1.6	1984, Kenya, Turkan Boy	900 cc brain size; Broca's area appears; est. 6 ft. tall; makes hand axe; min. body hair? Endured to 50 kya
18	*Homo heidelbergensis*	400 k	1992, Atapuerca, Spain	Discovered by Bermudez, Arsuaga, Carbonell

19	*Homo neanderthalensis*	350 k	1856, Germany	Steinheim Skull; powerful build; 5 to 6 ft. tall; archaic Homo sapiens
20	*Homo sapiens*	190 k	1974, Ethiopia	Omo 1 remains; EMH
21	*Homo sapiens idaltu*	160 k	2003, Ethiopia	Herto remains; AHS
22	*Homo floresiensis*	95 k	2003, Indonesia	LB 1 (Hobbit); immature design; dwarf-sized; 400 cc brain; stone tools
23	*Homo sapiens*	35 k	1868, France	Cro-Magnon I; EMH; 1,600 cc brain size
24	*Homo sapiens sapiens*	12 k	1920s onward, various locations: Egypt, Australia, USA, Algeria, Sudan, Indonesia, Mexico, Chile, Austria, England.	1400 cc brain; extra "sapiens" added by some taxonomists to indicate evidence of modern human behavior; AMH
25	Experimental	N/A	Numerous locations.	Design upgrade; 1500 cc brain

Appendix D

Taxonomy of Bipedal Creatures In Current Usage

Below is an abridged view of the archaic system used in 2012 by which fossils of bipedal creatures found worldwide are categorized. Notice how newer "improved" creatures are found to have larger brains. The fossils are arranged by approximate age according to radiometric and/or incremental dating. Species names are in accordance with current consensus.

Note: Glacial periods began 1.8 MYA. There have been nine glacial periods since they first began. The earth's magnetic field reversed mysteriously 780 KYA.

A = australopithicus = South African ape
H = homo = human

MYA	Brain Size (cc)	Taxonomic Category	Body Features	Tools
3–1	450–500	A. robustus	Ape-like, heavy, squat build, massive skull	Very crude, "pebble" stone tools.
3.0–1.5	450–550	A. africanus	Slim, flexible, ape-like omnivores	Very crude
2.0–1.0	500–550	A. robustus	Large back teeth, small front teeth	Crude choppers and scraping tools.
2.0–1.7	600–670	H. habilis	More ape-like than human, 4'3" tall, 80 lbs., hands good for gripping, opposed thumbs, longer arms for climbing trees, scavenger	Crude
2.0-0.5	775-1300	H.erectus	Better suited for walking, 5'6" tall, smaller back teeth, bony ridges over eyes, modern looking body, hands capable of "precision grip."	Acheulian hand axes for butchering large animals; tamed fire
400–300 KYA	1,000	H.erectus	Broca's area of brain enlarged, more capacity for language	Ditto
200–35 KYA	1,500	Neanderthal	Robust build	Stone tools, clubs, spears, composite tools

200 KYA	1,400	Early archaic man	Heavy build, massive brows	Better, more variety
120 KYA	1,400	Late archaic man	Mix of early and late features	Wide variety of stone tools
115 KYA	1,400	Anatomically modern	Gracile; only a few archaic features	
45–21 KYA	1,400	Cro-magnon	Round-headed, modern features	
12 KYA	1,400	Modern man	Modern features	Range from bow and arrow, woven baskets, flint-bladed sickles, etc., to computers, GPS, cell phones, etc.